Blood: Rheology, Hemolysis, Gas and Surface Interactions

Advances in Cardiovascular Physics

Vol. 3

Series Editor
D. N. Ghista, Houghton, Mich.

S. Karger · Basel · München · Paris · London · New York · Sydney

Blood: Rheology, Hemolysis, Gas and Surface Interactions

Volume Editors
D. N. Ghista, Houghton, Mich.; E. Van Vollenhoven, Delft;
W.-J. Yang, Ann Arbor, Mich., and H. Reul, Aachen

65 figures and 9 tables, 1979

S. Karger · Basel · München · Paris · London · New York · Sydney

Advances in Cardiovascular Physics

Vol. 1: Theoretical Foundations of Cardiovascular Processes. Eds. D. N. Ghista (Houghton, Mich.); E. Van Vollenhoven (Delft); W.-J. Yang (Ann Arbor, Mich.); H. Reul (Aachen). X + 182 p., 65 fig., 6 tab., 1979
ISBN 3-8055-2850-7

Vol. 2: Cardiograms: Theory and Applications. Eds. D. N. Ghista (Houghton, Mich.); E. Van Vollenhoven (Delft); W.-J. Yang (Ann Arbor, Mich.); H. Reul (Aachen). XII + 160 p., 79 fig., 4 tab., 1979
ISBN 3-8055-2851-5

National Library of Medicine Cataloging in Publication
  Blood: rheology, hemolysis, gas, and surface interactions
  Volume editors, D. N. Ghista ... [et al.]. – Basel, New York, Karger, 1979.
  (Advances in cardiovascular physics; v. 3)
  1. Blood – physiology I. Ghista, Dhanjoo N., ed. II. Series
  W1 AD53F v.3/WG 106 B652
  ISBN 3-8055-2852-3

All rights reserved.
No part of this publication may be translated into other languages, reproduced or utilized in any form or by any means, electronic or mechanical, including photocopying, recording, microcopying, or by any information storage and retrieval system, without permission in writing from the publisher.

© Copyright 1979 by S. Karger AG, 4011 Basel (Switzerland), Arnold-Böcklin-Strasse 25
Printed in Switzerland by Thür AG Offsetdruck, Pratteln
ISBN 3-8055-2852-3

Contents

Foreword IX
Preface X

Determination of Blood Rheological Parameters and Clinical Application

H. CHMIEL, Stuttgart 1

Abstract 1
Glossary/Terminology 2
List of Symbols 2
  I. Introduction 3
 II. Introduction to the Rheology of Viscoelastic Fluids 3
III. Blood Plasma and Blood as Viscoelastic Fluids 7
    A. The Rheological Behaviour of Normal Blood Plasma 7
    B. The Rheological Behaviour of Normal Human Blood 11
 IV. Blood Rheology in Clinical Pathology and Medicine 24
    A. Viscometer for Clinical Application 24
    B. Standard Values for Blood and Plasma 29
    C. Rheological Alterations of Blood Immediately after Heart Infarction 30
    D. The Influences of Medications on Blood Viscosity 34
    E. Blood Viscosity – Related Risk Factors of Heart Infarction ... 35
    F. Rheumatic Diseases 37
    G. Plasmocytoma 37
    H. Von Willebrand-Jürgens Syndrome 38
    I. Anaemias 38
    J. Blood Sedimentation Rate and Plasma Viscosity 39
    K. The Viscoelasticity of Blood for Patients with Peripheral Vascular Diseases and for Heavy Smokers 39
  V. Conclusion 40
References 41

Blood-Gas Interactions and Physiological Implications
W.J. YANG, Ann Arbor, Mich. 45

Abstract 46
List of Symbols 46
I. Introduction 48
II. Gas Exchange in the Lungs and Its Transport by the Blood 49
   A. The Lungs and Gas Exchange 49
      1. Composition of Atmospheric and Alveolar Airs 49
      2. Diffusion of Oxygen and Carbon Dioxide through the Alveolar-Capillary Membrane 50
      3. Rate of Gas Uptake in Lung Capillary Blood 52
   B. The Carriage of Oxygen and Carbon Dioxide in the Blood 55
III. Gas Emboli in Human Body 57
   A. Sources of Gas Emboli 57
      1. Accidentally Introduced 57
      2. Purposely Introduced 58
   B. Expansion and Dissolution of Gas Emboli in Blood 59
      1. Theory 59
         a) The Case Where Mass Diffusion Controls 59
         b) The Case Where Liquid Inertia Controls 65
      2. Experiments 67
   C. Effects of Foreign Agents on the Behavior of Gas Emboli 70
      1. Plasma Substitutes 70
      2. Anesthetics 71
IV. Gas Embolism Due to Extracorporeal Oxygenation 72
   A. Artificial Heart-Lung Machines 72
   B. Blood Trauma 73
      1. Possible Factors Responsible for Hemolysis 73
      2. Denaturation 74
   C. Introduction of Gas Emboli into the Blood during Open-Heart Surgery ... 74
   D. Hydrodynamic Basis of Hemolysis 75
      1. Sphering of Erythrocytes 75
      2. Critical Membrane Yield Stress 75
      3. Wall and RBC Collisions 76
      4. Shock Waves and Liquid Jets Produced by Collapsing Emboli 76
   E. Gas Embolism Syndromes, Prevention and Treatment 77
      1. Types of Air Embolism 77
      2. Treatment and Prophylactic Measures 79
V. Gas Embolism Due to Sudden Decompression 79
   A. Bubble Formation Due to Sudden Decompression 80
      1. Nucleation 80
      2. Stable Nuclei 81
      3. Criteria of Cavitation 81
      4. Factors Affecting Bubble Growth or Shrinkage 82
   B. Symptoms of Decompression Sickness 82
   C. Prevention and Treatment of Decompression Sickness 83
VI. Tissue-Capillary Gas Exchange 84
   A. Exchanges of Oxygen and Carbon Dioxide in the Tissues 85

Contents

```
    1. Exchange of Oxygen ............................................... 85
    2. Exchange of Carbon Dioxide ....................................... 86
  B. Dissolution of Gas Emboli in the Tissues ............................. 86
    1. In vivo Tests ..................................................... 87
    2. Theory ........................................................... 91
      a) Tissue Creep Controlling ....................................... 92
      b) Mass Transfer Controlling ...................................... 93
      c) Both Tissue Creep and Mass Transfer Controlling ................ 96
    3. Comparison of Theory with in vivo Tests .......................... 97
References ............................................................... 97
```

Biomaterials and Interfacial Phenomena

J. FEIJEN; T. BEUGELING; A. BANTJES, and C. TH. SMIT SIBINGA, Enschede/Groningen   100

```
Abstract .................................................................. 100
  I. Introduction ......................................................... 101
 II. Characterization of Surface and Interface ............................ 101
III. Correlation of Interfacial and Surface Parameters of Materials with Phe-
     nomena Occurring after Contact of These Materials with Blood or Plasma .. 107
 IV. Protein Adsorption onto Foreign Surfaces ............................. 108
     A. Type of Protein Adsorption and Measuring Techniques ............... 109
     B. Protein-Interface Interactions .................................... 110
        1. Apolar Surfaces ................................................ 111
        2. Polar Surfaces ................................................. 111
  V. Platelet Adhesion, Aggregation, and the Clotting of Blood ............ 115
     A. Platelet Adhesion and Aggregation ................................. 115
     B. Blood Coagulation ................................................. 116
     C. Foreign Materials ................................................. 119
 VI. Concluding Remarks .................................................. 126
Acknowledgements ......................................................... 126
References ............................................................... 127
```

Blood-Surface Interactions as a Basis for Selection of Blood-Compatible Cardiovascular Implantable Materials

S. SRINIVASAN; N. RAMASAMY; B. STANCZEWSKI, and P. N. SAWYER, Brooklyn, N.Y.   133

```
Abstract .................................................................. 134
Glossary .................................................................. 134
  I. Introduction ......................................................... 135
 II. Some Essential Criteria for the Selection of Blood-Compatible Materials .... 135
III. Correlations Between Interfacial Electrochemical Properties and Blood
     Compatibility of Materials ........................................... 136
     A. Conducting Materials .............................................. 136
     B. Insulator Materials ............................................... 143
 IV. Some Recent Techniques for Improving Blood Compatibility ............. 148
     A. Biolized Materials ................................................ 148
```

    B. Linear or Cross-Linked Homo or Block Polymers 149
    C. Surface Pretreatment of Metallic Materials 149
V. Blood-Surface Interactions – Mechanism Studies 150
    A. General 150
    B. Effect of Metals on the Platelet Release Reaction 151
    C. Electron Microscopic and Optical Studies of Surfaces Exposed to Biologic Fluids 152
    D. Electrochemical Reactions of Blood Coagulation Factors 154
    E. Adsorption and Adsorption Inhibition 158
VI. Conclusions 159
Acknowledgements 160
References 160

    Subject Index 164

Foreword

The theme of the 'Advances in Cardiovascular Physics' series is to highlight the foundations of cardiovascular mechanisms and its clinical procedures and devices. In keeping with this theme, this volume deals with the techniques of using blood rheology in following the course of diseases, mechanisms of and safeguards against blood gas embolism, and the criteria and selection of blood-compatible prosthetic materials.

The volumes of this series cater to the job or professional descriptions of clinical engineers and physicists. The volumes are also intended to provide the clinicians insight into the (engineering physics) basis of physiological syndromes that they treat and the methods and devices that they use in clinical practice. Of course, these volumes are particularly prepared to make them suitable for use as course texts and references. In fact, with this particular purpose in mind, Volume I has been prepared to uniquely provide the governing theory and equations of cardiovascular electro-magnetic and transport processes. The subsequent volumes are designed to bridge the gap from cardiovascular theory to practice.

DHANJOO N. GHISTA

Preface

Blood is the single most important element of the cardiovascular system. While its rheology and gas interactions have important clinical significance, its interactions with artificial surfaces of assist-prosthetic and extracorporeal devices contribute to their efficacies. This book brings together the underlying mechanisms of blood rheology and (gas and surface) interactions, methods of quantifying them, and their clinical significances.

In the first chapter, the rheological behavior of healthy and pathological human blood is outlined. Also presented are (i) the techniques for measurement of blood rheology, and (ii) its use as a diagnostic tool in clinical pathology and medicine, and as a means for observing the course of diseases. The chapter on blood gas interactions deals with the physics of gas exchange, kinetics and circumstances of gas embolism syndrome (in extracorporeal oxygenation and during decompression) and its treatment.

The remaining two chapters are on blood-surface interactions. The influences of electrochemical characteristics across interfaces and of the surface parameters on the interfacial interactions are delineated, in order to provide insight into the mechanisms and criteria of blood compatibility of prosthetic materials. Techniques for determining the interactions (for evaluation of prosthetic materials) are presented, along with the results of implementation on valves and tubes of conducting and insulating materials. Methods of improving blood compatibility of prosthetic material candidates are presented.

The utility of this rigorous cum state-of-the-art cum instructional book is that it brings together, under one cover, the mechanisms and clinical implications of all aspects of blood interactions, relevant to physiological function and compatibility of prosthetic-assist devices.

Dhanjoo N. Ghista   Wen-Jei Yang
Eric Van Vollenhoven   Helmut Reul

Adv. cardiovasc. Phys., vol. 3, pp. 1–44 (Karger, Basel 1979)

Determination of Blood Rheological Parameters and Clinical Application

H. CHMIEL

Institut für Grenzflächen- und Bioverfahrenstechnik, Stuttgart

Contents

Abstract	1
Glossary/Terminology	2
List of Symbols	2
I. Introduction	3
II. Introduction to the Rheology of Viscoelastic Fluids	3
III. Blood Plasma and Blood as Viscoelastic Fluids	7
A. The Rheological Behaviour of Normal Blood Plasma	7
B. The Rheological Behaviour of Normal Human Blood	11
IV. Blood Rheology in Clinical Pathology and Medicine	24
A. Viscometer for Clinical Application	24
B. Standard Values for Blood and Plasma	29
C. Rheological Alterations of Blood Immediately after Heart Infarction	30
D. The Influences of Medications on Blood Viscosity	34
E. Blood Viscosity – Related Risk Factors of Heart Infarction	35
F. Rheumatic Diseases	37
G. Plasmocytoma	37
H. Von Willebrand-Jürgens Syndrome	38
I. Anaemias	38
J. Blood Sedimentation Rate and Plasma Viscosity	39
K. The Viscoelasticity of Blood for Patients with Peripheral Vascular Diseases and for Heavy Smokers	39
V. Conclusion	40
References	41

Abstract

The rheological formulation of human blood is discussed. It is shown that human blood behaves like a viscoelastic fluid. At very low shear rates, the viscosity of blood becomes constant, which means that the yield shear stress does not exist.

A few new viscometers suitable for clinical application are introduced. They allow one to distinguish between an increased tendency of the erythrocytes to aggregate and a decrease in their flexibility, as causes for observed increased viscosity of blood.

For diagnostic purposes, normal values of plasma and blood viscosities at selected shear rates are presented. The clinical significance of changes in the viscosity of blood and plasma is demonstrated using as examples, heart infarction and its risk factors, rheumatic diseases, plasmocytoma, the von Willebrand-Jürgens syndrome and anaemias.

Glossary/Terminology

(1) *Shear rate:* Tidal change of deformation of a fluid element in plane laminar flow.

(2) *Shear stress:* Force per unit of fluid element area when this element is exposed to a certain shear rate.

(3) *Non-Newtonian fluids:* Fluids whose flow behaviour cannot be described by the Navier-Stokes equations.

(4) *Viscoelastic fluids:* While for Newtonian fluids the normal stresses are identical in all directions, viscoelastic fluids have anisotropic stress distributions.

(5) *Relaxation time:* Besides the anisotropic behaviour, viscoelastic fluids show a tidal shifting between a rapid change of deformation and the adjusting equilibrium stress state.

(6) *Yield shear stress:* Necessary minimum shear stress to generate a flow in a certain fluid.

(7) *Erythrocyte flexibility:* Measure for the deformability of erythrocytes.

(8) *Erythrocyte-aggregation tendency:* Tendency of erythrocytes to form rouleaux-type aggregations at low shear rates.

(9) *Fahraeus-Lindquist effect:* Separation processes within blood flow near walls. First observed in capillary blood flow. They are, however, to be found in any shear flow (in cone-plate or Couette-system as well).

List of Symbols

Symbol	Description	Symbol	Description
A, m^2	area	$K = B/P$	
a, Nm^{-2}	constant	$\Delta L, m$	length difference
$B = \eta_B/\eta_N$	viscosity ratio for blood	M, Nm	torque
b	constant of the Casson equation	n, sec^{-1}	rotational speed
		$P = \eta_p/\eta_N$	viscosity ratio for plasma
c, sec^{-1}	constant	p, Nm^{-2}	pressure
D, m	pipe diameter, sphere diameter	$-p^+, Nm^{-2}$	isotropic part of the normal stress
d	constant	$\Delta p, Nm^{-2}$	pressure difference
e	constant	p', Nm^{-3}	pressure gradient
f, Hz	frequency	$Q, m^3 sec^{-1}$	flow
F, N	force	R, m	sphere radius, tube radius
F_1, F_2	normal-stress functions	r, m	radial coordinate
$g, °C$	constant	$t, °C$	temperature
$h, °C$	constant	t_0^*, sec	relaxation time
H, m	height of measuring cylinder	$t_0, °C$	room temperature (23 °C)

Symbol	Description	Symbol	Description
v, m sec^{-1}	velocity	η' Nsec m^{-2}	viscous component of complex viscosity
$\bar{v}$, m sec^{-1}	average velocity		
x, y, z	coordinates	η'', Nsec m^{-2}	elastic component of complex viscosity
$\beta = (R_i/R_a)^2$	ratio of radii of inner and outer cylinders (Couette)	η_∞, Nsec m^{-2}	viscosity for 'infinitely' high shear rate
$\dot{\gamma}$, sec^{-1}	shear rate		
$\hat{\gamma}$, sec^{-1}	representative shear rate (at the radial distance $\frac{\pi}{4}R$)	θ	phase angle of complex viscosity
		ρ, kg m^{-3}	density
γ	shear angle	σ, Nm^{-2}	normal stress
$\dot{\gamma}_w$, sec^{-1}	wall shear rate	τ, Nm^{-2}	shear stress
$\dot{\gamma}_w^*$	complex wall shear rate	τ_0, Nm^{-2}	yield shear stress
η, N sec m^{-2}	viscosity	τ_w^*	complex wall shear stress
η_B, Nsec m^{-2}	measured blood viscosity	$\dot{\tau}$, Nm^{-2} sec^{-1}	rate of variation of shear stress with time
η_N, Nsec m^{-2}	standard blood viscosity		
η_N, Nsec m^{-2}	standard plasma viscosity	$\Omega\,\omega$, sec^{-1}	angular velocities
η_p, Nsec m^{-2}	measured plasma viscosity	*Indices*	
η_0, Nsec m^{-2}	zero viscosity (viscosity at zero shear rate)	a	outer cylinder
		i	inner cylinder
η^*	complex viscosity	w	conditions on the wall

I. Introduction

Investigation of blood rheology is by no means so recent as one would suppose. POISEUILLE, in 1846, had already ascertained that blood does not behave like 'normal' fluids. In fact, blood is a typical viscoelastic fluid [THURSTON, 1972; WELLS and MERRILL, 1961]. Within the last 10 years, the interest in blood rheology – in particular in the study of microcirculation disturbances – has greatly increased. Recent studies by CHMIEL and STÖRMER [1972], Störmer et al. [1973b] and BOSS et al. [1973] have shown that in most cases a microcirculation disturbance is accompanied by a pathological change in certain rheological blood parameters.

In this chapter, the rheological formulation of blood is presented, and the rheological behaviours of healthy and pathological human blood are discussed. Further, the experimental determination of the rheological parameters is indicated, and the clinical importance of blood rheology is elucidated.

II. Introduction to the Rheology of Viscoelastic Fluids

Consider two parallel plates with a fluid layer between them. One plate is moved in relation to the other at a constant speed of v_0. Assuming that the fluid adheres to the walls, a linear velocity profile occurs, in the steady state, due to internal forces in the fluid layer (with the exception of the area at the plate edges), as shown in figure 1. That is to say, the flow rate steadily decreases from v_0 on the moving plate to zero on the fixed plate. The force per unit area of an element in the fluid, $\Delta F/\Delta A$, caused by the movement of the plate, is (i) directly proportional to the velocity difference Δv across the element (of thickness Δy) and its dynamic viscosity, and (ii) inversely proportional to the thickness of the element. Hence,

$$\frac{\Delta F}{\Delta A} = \eta \frac{\Delta v}{\Delta y}.$$

If the fluid element becomes vanishingly small and one substitutes the shear stress, τ, for $\Delta F/\Delta A$, and the velocity gradient dv/dy for $\Delta v/\Delta y$, one obtains:

$$\tau = \eta \frac{dv}{dy}. \tag{1}$$

The fluid element is deformed due to the shear stress, as shown in figure 2. It can be seen from the illustration that, in the case shown here, the velocity gradient dv/dy is identical with the continuous change in the degree of deformation, γ, i.e., $d\gamma/dt$. The latter is termed shear rate and is abbreviated as $\dot{\gamma}$. Equation 1 can, thereby, be written simply as:

$$\tau = \eta \dot{\gamma}. \tag{2}$$

In a great number of fluids (as for example polymer solutions, suspensions of flexible particles and, therefore, blood itself) the viscosity is not a constant of the material, but decreases with increasing shear stress (shear rate). This is illustrated in figure 3, using an aqueous polymer solution. Fluids in which the viscosity is a function of the shear rate are described as non-Newtonian.[1] Their viscosity is no longer a material-specific constant, but is dependent on the stress. Equation 2 is used here to define 'variable viscosity':

$$\eta = \frac{\tau}{\dot{\gamma}}.$$

---

[1] The flow curve is, however, not sufficiently comprehensive to be used as a criterion for non-Newtonian fluids. To be more precise, all fluids whose flow behaviour cannot be described by the basic equations of classical hydrodynamics (Navier-Stokes equations) should be regarded as non-Newtonian.

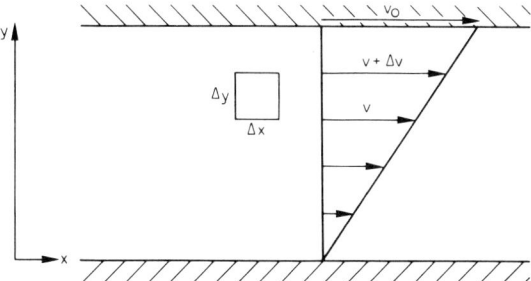

Fig. 1. Velocity distribution in steady state two-dimensional laminar flow.

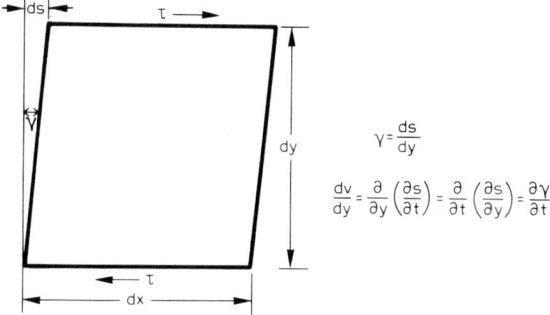

Fig. 2. Deformation of an infinitesimal fluid element for laminar flow between parallel surfaces.

The shape of the viscosity versus shear rate curve, as shown in figure 3, is typical of the group of fluids known in general rheology as viscoelastic fluids. The initial viscosity value at infinitesimal stress is also called the zero viscosity, η_0. It remains constant for a specific range of shear rate, within the limits of experimental precision. With increasing stress, the viscosity is greatly reduced. The curve goes through a point of inflexion, and at very high shear rates approaches a lower limit ($\eta \approx \eta_\infty$). In addition to the variable viscosity, the above-mentioned fluids can be rheologically distinguished from Newtonian fluids by the occurrence of elastic effects, e.g., they show characteristic relaxation times.

The stresses on a fluid element due to laminar flow are depicted spatially in figure 4. In addition to the flow direction marked x and the shear direction marked y, a further coordinate is necessary – pointing in the passive direction – which is marked z. It is well known that the normal stresses, σ, which occur in the three directions, in a Newtonian fluid are equal, i.e.,

$$\sigma_x = \sigma_y = \sigma_z = -p \tag{3}$$

From experience, the three normal stresses in viscoelastic fluids are not equal. If one assumes that the stress in the z-direction is $-p^+$, then one may describe the rheological behaviour of viscoelastic fluids in the particular case shown here, i.e., laminar flow between parallel surfaces, by

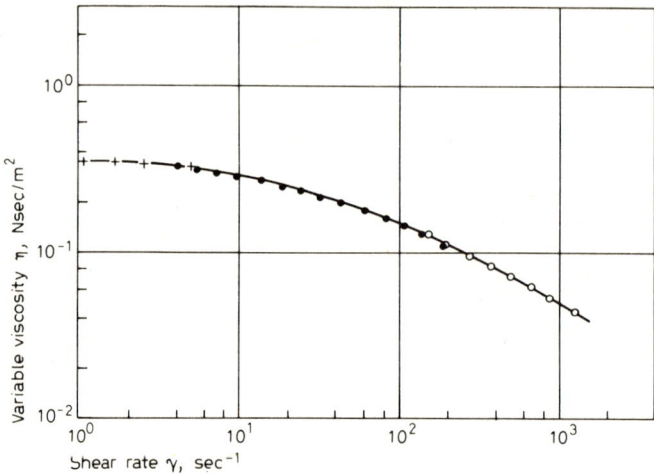

Fig. 3. Variable viscosity as function of shear rate, for a polymer solution (0.6% carboxymethylcellulose in water). ○ = Cone-plate viscometer; ● = Couette viscometer with narrow gap; + = Couette viscometer with wide gap.

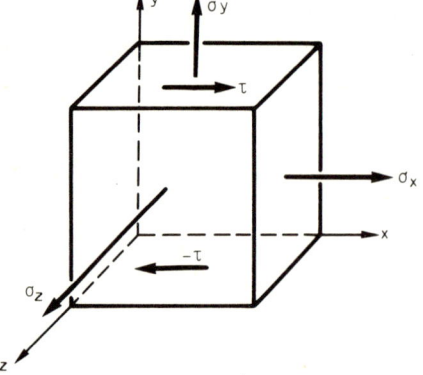

Fig. 4. Stresses on a volume element in a fluid for laminar flow between parallel surfaces.

the following functions:

$$\tau = f(\dot{\gamma})$$
$$\sigma_x = -p^+ + F_1$$
$$\sigma_y = -p^+ + F_2 \quad (4)$$
$$\sigma_z = -p^+.$$

The isotropic component $-p^+$, which occurs in the three normal stresses, is generally dependent on $\dot{\gamma}$, but is identical with the hydrostatic pressure for $\dot{\gamma} = 0$. Both the normal stress functions F_1 and F_2 are determined, as one can see, by taking the difference of the normal stresses, i.e., $F_1 = \sigma_x - \sigma_z$ and $F_2 = \sigma_y - \sigma_z$. Because the second normal stress function F_2 is found experimentally to be very small, it shall be neglected here. In this way, the parameters of the 'zero stress' characteristic of viscoelastic fluids can be defined:

$$\eta_0 = \lim_{\dot{\gamma} \to 0} \frac{\tau}{\dot{\gamma}}; \quad t_0^* = \frac{1}{2\eta_0} \lim_{\dot{\gamma} \to 0} \frac{F_1}{\dot{\gamma}^2} \quad (5)$$

where t_0^* is the relaxation time of the fluid in question.

The experimental determination of the flow curve $f(\dot{\gamma})$ and relaxation time t_0^* will be discussed, in connection with blood rheology, in the next section. It should be mentioned here, however, that most of the rheometers used in general rheology are unsuitable for blood, for reasons which will be explained.

III. Blood Plasma and Blood as Viscoelastic Fluids

A. The Rheological Behaviour of Normal Blood Plasma

As is well known, blood plasma consists (if we ignore the inorganic components) of proteins (whose molecular weight in part exceeds 10^6) which are dissolved in water. It could be concluded, from this, that blood plasma behaves rheologically like the polymer solution illustrated in figure 3. In earlier works, as for example WELLS and MERRILL [1961], plasma viscosity is clearly described as being dependent on shear rate. On the other hand, in later publications, plasma is described as a Newtonian fluid. Very recent articles by COPLEY [1971], and COPLEY and KING [1972] attempt to explain this contradiction. According to these studies, the dependence of the viscosity of blood plasma on the shear stress is due to a systematic error whose cause is to be found in a surface effect at the phase boundary plasma/air. COPLEY and his colleagues state that in the presence of fibrinogen and other high molecular weight proteins, a 'polymolecular' layer is formed at the liquid surface which stimulates a yield shear stress in the plasma.

Formerly, plasma viscosity was measured chiefly in the so-called Couette viscometer. This viscometer consists of two concentrically arranged cylinders, one of which is set in rotation with an angular velocity of Ω, while the other remains stationary. The fluid to be tested is introduced into the annular space between the inner and outer cylinders. The torque, which is necessary to hold the stationary cylinder in position, is a measure of the viscosity of the fluid. The theoretical considerations for the calculation of shear stress and the related shear rate from the measured data, angular velocity, Ω, and torque, M, have already been fully dealt with in the literature on the narrow Couette gap, for both Newtonian and non-Newtonian fluids [e.g. by OKA, 1960]. The derivation will, therefore, be omitted. The stress and shear rate on the inner cylinder are given by

$$\tau(R_i) = \frac{M}{2\pi H R_i^2} \tag{6}$$

and

$$\dot{\gamma}(R_i) = \frac{2\Omega}{(R_a/R_i)^2 - 1} \tag{7}$$

where R_a is the radius of the outer cylinder, R_i the radius of the inner cylinder, and H the effective height of the cylinder. Corresponding pairs of τ and $\dot{\gamma}$ values can be employed to obtain a plot of τ versus $\dot{\gamma}$, the flow curve.

If the flow curve of normal human plasma is determined using the Couette system (illustrated in fig. 5a), the apparent yield shear stress is observed (as already mentioned), as a result of the surface effect. If, however, one uses the 'guard ring' [suggested by COPLEY and KING, 1972], shown in figure 5b or in the modified system illustrated in figure 5c, the error which then occurs due to this effect is within the bounds of experimental error. The method is not entirely satisfactory, however. The author, therefore, has developed in addition to a new form of capillary viscometer (which will be fully described in the next section), a sphere-sphere system [CHMIEL, 1974b], which is particularly suitable for the viscosity measurement of blood and plasma (fig. 6).

The derivation of the formula for the calculation of shear rate and viscosity in the narrow gap between the spheres will not be dealt with here. The reader is referred to the works by OKA [1960] and v. BRACHEL and SCHÜMMER [1975]. For a sphere of outer radius R_a, one obtains, at an

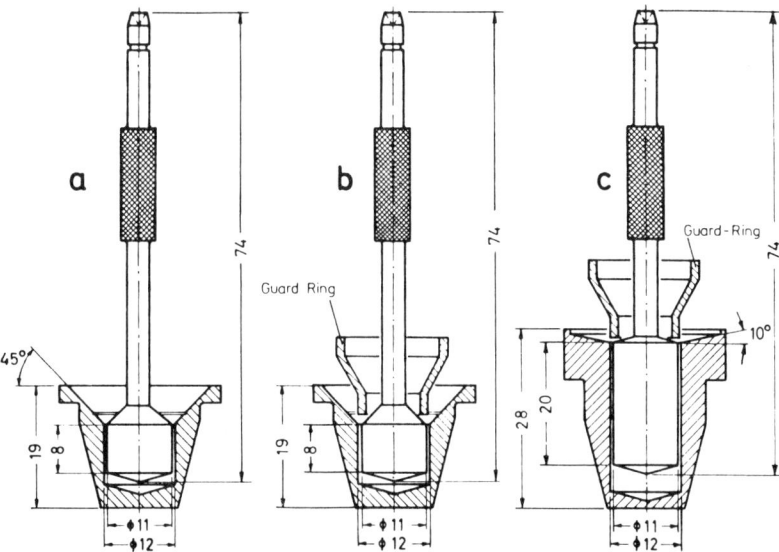

Fig. 5. Couette measuring system with a narrow gap (scale in mm). The outer cylinder is rotated, and the moment to prevent rotation of the inner cylinder is measured. *a* Commercially obtainable model. *b.* Same system with a guard ring. *c* System developed to reduce surface effect to a minimum.

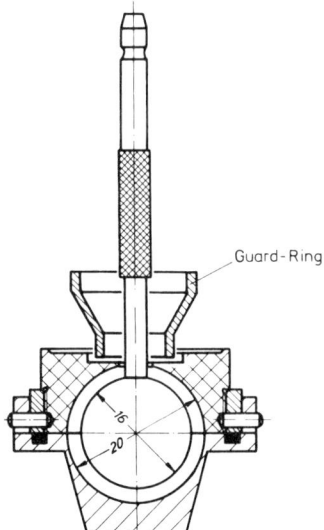

Fig. 6. Sphere-sphere measuring system, for minimising the surface effects which occur during the viscosity measurement of protein-containing fluids.

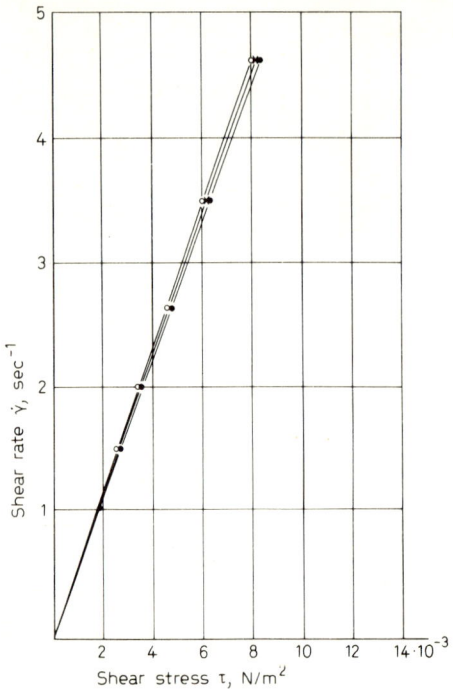

Fig. 7. Flow curve of normal human plasma, measured in the sphere-sphere viscometer. The results are from 3 different donors.

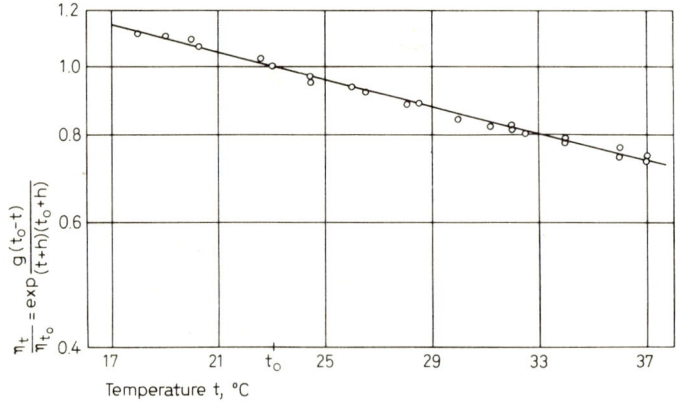

Fig. 8. Temperature dependence of plasma viscosity. $t_0 = 23\,°C$; $g = 420\,°C$; $h = 111\,°C$.

inner sphere of radius R_i, a measured torque M at a constant angular velocity, Ω. Hence

$$\dot{\gamma} = \frac{2.7 \cdot \Omega}{\pi(1 - R_i/R_a)} \tag{8}$$

and

$$\eta = \frac{\left(\dfrac{1}{R_i^3} - \dfrac{1}{R_a^3}\right) M}{8\pi\Omega}. \tag{9}$$

For non-Newtonian liquids, however, equations 8 and 9 are only valid with sufficient accuracy for $R_i/R_a \geq 0.8$, which condition is fullfilled in the case of the sphere-sphere viscometer used (fig. 6).

Figure 7 shows the flow curve for healthy human plasma measured using the sphere-sphere viscometer illustrated in figure 6.[2] It is clear that the measured points lie on a straight line, which passes through the origin. It can therefore be concluded that, using this measuring system, the surface effect, which is constantly present in plasma, is suppressed within the limits of experimental accuracy, and further that the viscosity is practically independent of shear rate.

According to equation 9, the η value obtained is $1.75 \cdot 10^{-3}$ Nsec/m² (1.75 cP) at 23 °C. However, as previous measurements by CHMIEL [1974a] have shown, plasma may not be described as a Newtonian fluid, as it shows typical viscoelastic effects, such as drag reduction in the case of turbulent flow. The temperature dependence of the viscosity of healthy human plasma at constant $\dot{\gamma}$ is illustrated in figure 8. It is practically the same as that of water and may be mathematically described by the following equation:

$$\eta_t = \eta_{t_0} \exp \frac{g(t_0 - t)}{(t + h)(t_0 + h)} \tag{10}$$

where t is the temperature of the measurement, t_0 is the room temperature (23 °C) and g and h are constants.

B. The Rheological Behaviour of Normal Human Blood

As is well known, blood is a suspension of highly flexible, corpuscular constituents (namely, erythrocytes, leucocytes and thrombocytes) in an aqueous polymer solution (blood plasma). One might, therefore, assume

---

[2] The drive and measurement unit of the 'LS 100' viscometer, manufactured by Contraves AG, Zürich, Switzerland, was used both here and in all other rotational viscometer measurements described in this chapter.

that it behaves rheologically like the viscoelastic fluids described in section II. A comprehensive study shows, however, that although blood is regarded as a non-Newtonian fluid, the shear dependence of the viscosity can be attributed to the formation or disintegration of erythrocyte aggregates [GOLDSTONE et al., 1970; SCHMIDT-SCHÖNBEIN, and WELLS, 1969, 1971]. These assertions are based chiefly on microscopic observations in transparent cone-plate viscometers [SCHMIDT-SCHÖNBEIN et al., 1969; SCHMIDT-SCHÖNBEIN and WELLS, 1971; CHIEN et al., 1967a], from which it is concluded that the erythrocyte aggregates also occur in normal human blood. The formation of aggregates is reversible and increases at decreasing shear rates.

It is also commonly held that blood possesses a yield shear stress [SCHMIDT-SCHÖNBEIN and WELLS, 1971; COKELET et al., 1963; CHARM, 1967; STOLTZ and LARCON, 1970]. An equation of the form

$$b\sqrt{\dot{\gamma}} = \sqrt{\tau} + \sqrt{\tau_0}, \tag{11}$$

which is known as the 'Casson equation', has been suggested to describe the flow curve of blood mathematically. Here τ_0 is the yield shear stress and b is a constant. SCOTT-BLAIR [1958] has already pointed out, in an earlier work, that this equation can only be applied to a small range of the rheology curve of blood. Furthermore, the values (for the same haematocrit) of τ_0 given in various works show considerable scattering. The explanation for this scattering is chiefly due to the hitherto minimal achieved shear rates of $\dot{\gamma} = 0.1\,\text{sec}^{-1}$ being regarded as extremely low. Extrapolation of the $\sqrt{\tau}$ versus $\sqrt{\dot{\gamma}}$ curve to $\dot{\gamma} = 0$ provides the value of τ_0. However, the value of τ_0, obtained in this manner (for a given haematocrit), is essentially dependent on the minimal shear rates that could be reproducibly measured.

In recent studies, CHMIEL [1973, 1974a] has investigated experimentally (a) whether blood possesses a yield shear stress and (b) whether the variable viscosity (shear dependence) of blood really does depend entirely on the formation and disintegration of erythrocyte aggregates. On the basis of their measurements, CHIEN et al. [1970] have, however, doubted the validity of both these assertions.

In order to answer these questions, it was necessary to achieve very low shear rates experimentally. The range from $\dot{\gamma} = 0.1$ to $4.6\,\text{sec}^{-1}$ was covered using the Couette system, illustrated in figure 5c. Figure 9 shows the flow curve of human blood with a haematocrit of 44% (vol% of the corpuscular constituents in the blood) obtained by this method. The same results were again plotted in the so-called Casson diagram (fig. 10). From both diagrams, it is clear that blood does not possess a yield shear stress.

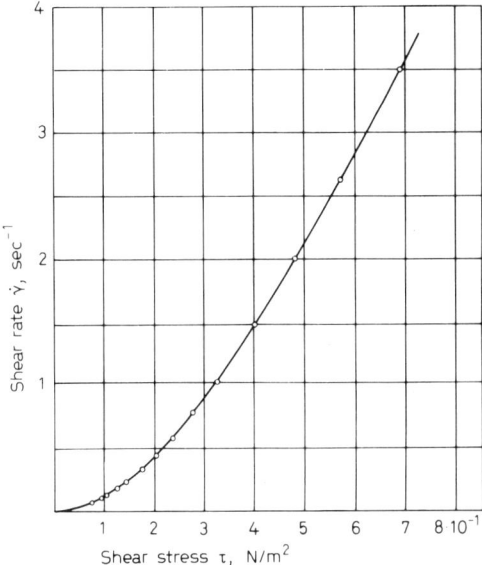

Fig. 9. Flow curve of normal human blood, with a haematocrit of 44%.

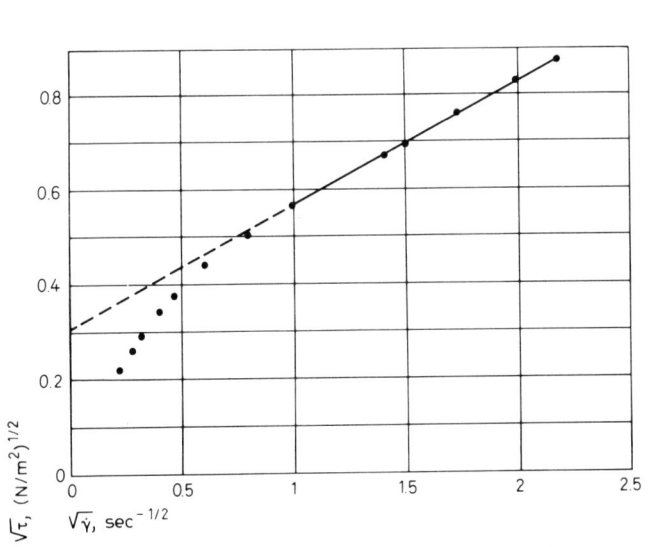

Fig. 10. 'Casson representation' of normal human blood, with a haematocrit of 44%. The straight line represents Casson's equation. It can be seen that the observed values deviate considerably for low values of $\sqrt{\dot{\gamma}}$, the deviation increasing at lower values of $\sqrt{\dot{\gamma}}$.

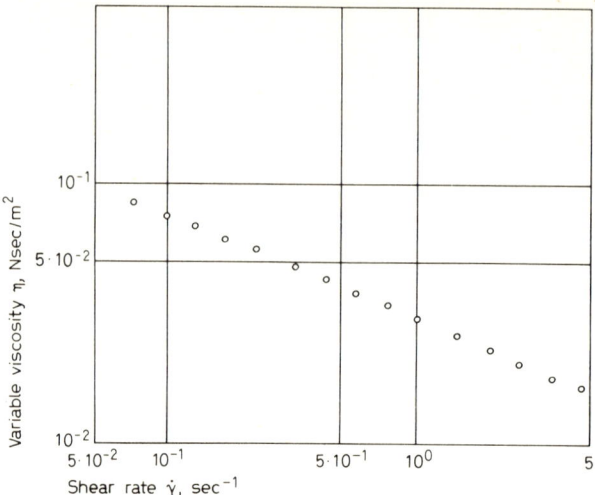

Fig. 11. Variable viscosity as function of shear rate for normal human blood, with a haematocrit of 44%.

As can be seen in figure 11, a representation in the form 'variable viscosity' as a function of shear rate does not permit an estimation of the zero viscosity (analogous to that in fig. 3). For this it would be necessary to realise even lower shear rates. This is possible using the Couette system, with the wide Couette gap, as shown in figure 12.

The theoretical considerations for the evaluation of measurements of non-Newtonian fluids, using this measuring system, were fully dealt with by CHMIEL and SCHÜMMER [1972] and CHMIEL [1975c]. For every measured value of the angular velocity Ω and of torque M, a corresponding pair of coordinates $(\tau, \dot\gamma)$ may be found using the following equations:

$$\dot\gamma(R_i) = 2\Omega \sqrt{\frac{1-\beta^3}{3(1-\beta)^3}} \qquad (12)$$

$$\tau = \frac{M}{2\pi H R_i^2} \sqrt{\frac{1-\beta^3}{3(1-\beta)}} \qquad (13)$$

where β is $(R_i/R_a)^2$.

In this way, one can, for example, determine shear rates of blood for values as low as $\dot\gamma = 5 \cdot 10^{-3}$ sec^{-1}. In figure 13 are shown the results of

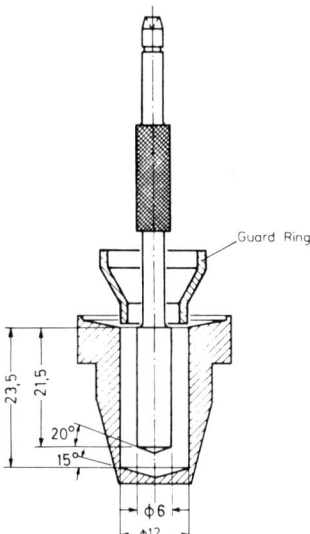

Fig. 12. Couette system with wide gap (scale in mm). This is used for extremely low values of $\dot{\gamma}$.

the measurements of blood viscosity in the wide Couette gap. It is quite obvious that with decreasing stress the variable viscosity tends towards a constant value (zero viscosity). The variation in the viscosity for $\dot{\gamma} \leq 10^{-2}\,\text{sec}^{-1}$ is within experimental accuracy. The question of the existence of a yield shear stress for blood can, therefore, be clearly answered in the negative.

Let us examine the assertion that the shear viscosity of blood is related exclusively to the aggregate formation of the erythrocytes. It was experimentally shown in a series of publications [SCHMIDT-SCHÖNBEIN and WELLS, 1971; CHIEN et al., 1967a] that aggregate formation by the erythrocytes only occurs in the presence of macro-molecules (e.g., fibrinogen, α_2-macroglobulin, etc.). This condition permits one to investigate the origin of the shear viscosity of blood. In these experiments, normal human blood was centrifuged and the surviving plasma removed. The erythrocytes were then washed in an isotonic salt solution (approx. 0.9% NaCl in H_2O). Finally, one half of the washed erythrocytes were used to produce suspensions of different concentrations, between 30 and 60 vol%, in isotonic NaCl. The other half of the centrifuged erythrocytes were 'hardened' in a 5.6% glutaraldehyde solution. After hardening for

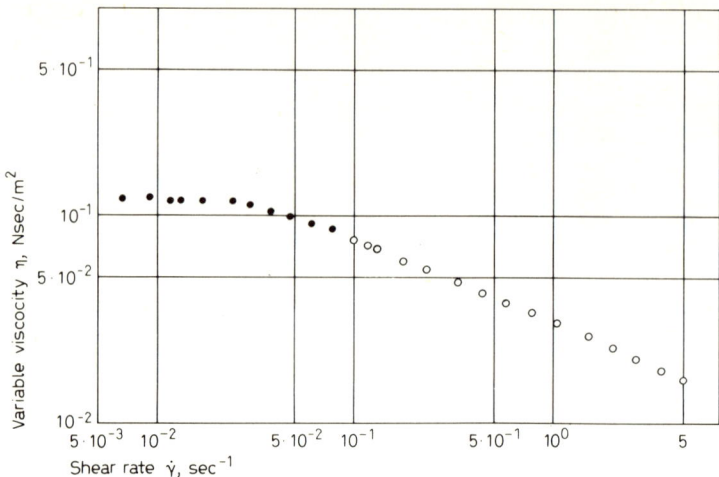

Fig. 13. Variable viscosity as function of shear rate for normal human blood with a haematocrit of 44%. Results are shown using the wide gap (●) and the narrow gap (○).

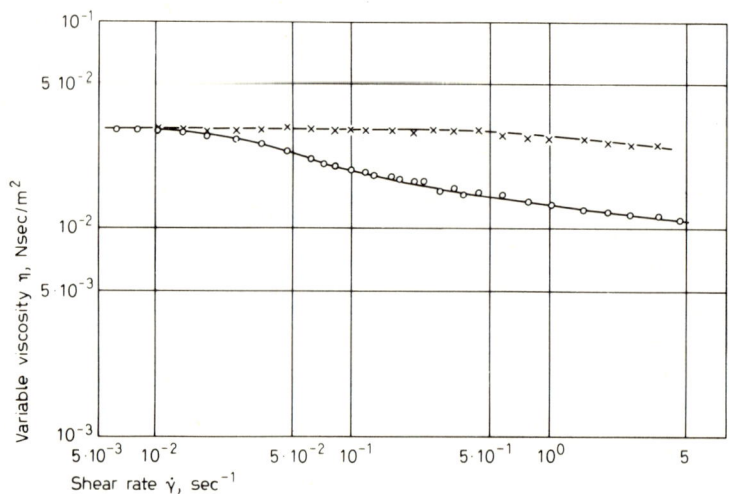

Fig. 14. Variable viscosity as function of shear rate for a 50 vol% suspension of normal (unhardened, ○) and hardened (×) human erythrocytes in an isotonic salt solution.

periods of 4, 24, 48 hrs, and 4 days, portions of the hardened erythrocytes were again washed and suspended in an isotonic salt solution in concentrations of between 30 and 55 vol%; the concentration was usually determined by micro-centrifugation, but also in some cases (as a check) using the Coulter counter.

It must be pointed out here that with normal unhardened erythrocytes, a packing density of 97–98 vol% can be attained (as is known from specialist literature), whereas for hardened erythrocytes it is approximately 60%. In the latter case, therefore, the haematocrit attained by means of micro-centrifugation must be multiplied by the factor 0.6 in order to calculate the actual suspension concentration in volume percent.

Assuming that the shear viscosity depends entirely on the formation of aggregates, a suspension of normal erythrocytes in isotonic salt solution must possess a linear flow curve. However, as can plainly be seen from figure 14, this is not so. On the contrary, the viscosity in the shear rate range shown here already varies by the factor 3. Comparison with the hardened erythrocytes in the same volume concentration is even more informative. Here is experimental proof of what was previously claimed in an earlier work [CHMIEL and STÖRMER, 1972], namely that the intrinsic rigidity of the individual erythrocyte causes it to behave rheologically, at low stresses, like a hardened erythrocyte. Within the accuracy of the measurement, both suspensions show the same zero viscosity. At higher shear rates, the influence of the intrinsic rigidity of the normal erythrocytes – and thereby their shear viscosity – decreases. At extremely high shear rates, the viscosity again tends towards a limit, corresponding to maximum deformation, which is defined here as η_∞. In normal blood, of course, the aggregation and break-up of aggregates of the erythrocytes have a great influence on the shear viscosity.

If one wants to calculate velocity profiles for parallel laminar flow of other geometries with the help of the experimental flow curve of blood, one requires a mathematical expression which approximates as closely as possible to the experimentally obtained data points. A 7-th degree polynomial of the following form is suitable:

$$\dot\gamma = c\left[\frac{\tau}{a} + \left(\frac{\tau}{a}\right)^3 + d\left(\frac{\tau}{a}\right)^5 + e\left(\frac{\tau}{a}\right)^7\right]. \tag{14}$$

The usefulness of this equation is illustrated in figure 15. Here, computed curves have been fitted to data points for three different flow curves corresponding to varying haematocrit, HK. The four constants a, c, d and e were determined from the data using the method of least squares. There is satisfactory agreement for the range of measurement shown, i.e., $\dot\gamma = 5 \cdot 10^{-3}$ to $\dot\gamma \approx 5\ \text{sec}^{-1}$.

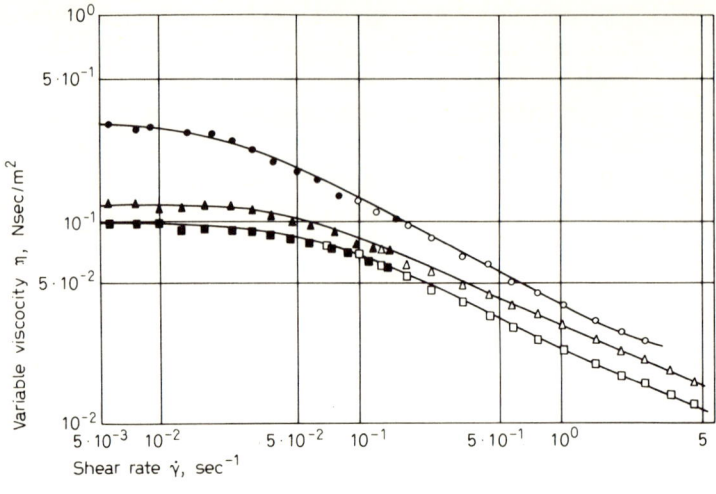

$$\dot{\gamma} = c\left[\frac{\tau}{a} + \left(\frac{\tau}{a}\right)^3 + d\left(\frac{\tau}{a}\right)^5 + e\left(\frac{\tau}{a}\right)^7\right].$$

HK %	LSW	LS	a N/m²	c sec⁻¹	d	e · 10⁺³
50	●	○	0.013	0.043	−0.0326	0.370
44	▲	△	0.0153	0.128	−0.071	1.716
40	■	□	0.0114	0.1156	−0.032	1.310

Fig. 15. Variable viscosity as function of shear rate for normal human blood, where the parameter is the haematocrit, HK. LSW indicates measurements made using a wide gap and LS, using a narrow gap.

The methods of measurement so far discussed permit only the lower range of shear rates (which is, however, the most important one for the erythrocyte aggregation) to be examined viscometrically. As will be shown in the next section, the higher shear rate range is also of interest for clinical tests, because, as can be seen from figure 14, it is particularly suitable for examining possible changes in the erythrocyte flexibility and its influence on viscosity.

Although one often reads about the use of cone-plate viscometers for the measurement of blood viscosity, they are entirely unsuitable for this purpose. Even a small sedimentation, which is always present in blood, leads to considerable experimental errors. The capillary viscometer is, on the other hand, highly suitable for this purpose. Among the many different types of flow, flow along tubes is most relevant to research not only for the natural circulation, but also for modelling blood flow in the apparatus for medical technology (e.g., membrane oxygenators). This topic has already been dealt with frequently in the literature [see, for example, COWIN, 1972: SKALAK et al., 1972; BUGLIARELLO and SEVILLA, 1970; HERSHEY and CHO, 1966; CHISHOLM and GAINER, 1971; GOLDSMITH, 1971].

What is incomprehensible, however, is that even though one finds that a non-linear law, e.g., the Casson law, is required to describe the flow curve, many authors still use the linear Hagen-Poiseuille law to calculate the shear rate of viscosity at the tube wall, without regard for the errors introduced in this way. Knowledge of the variable viscosity as a function of the shear rate for values of $\dot{\gamma}$, approaching zero, allows us to estimate in advance both the velocity profile and the shear rate distribution in laminar flow of blood through tubes. A comparison with the relative data for the Hagen-Poiseuille flow gives information about the errors that arise when using this law. For normal human blood with a haematocrit of 50%, whose variable viscosity is shown in figure 15, the method of least squares gave, for a 7-th degree polynomial (eq. 14), the following constants:

$a = 0.013$ N/m^2, $c = 0.043$ sec^{-1}, $d = -0.0326$, $e = 0.370 \cdot 10^{-3}$

Hence

$$\dot{\gamma} = 0.043\left[\left(\frac{\tau}{0.013}\right) + \left(\frac{\tau}{0.013}\right)^3 - 0.0326\left(\frac{\tau}{0.013}\right)^5 + 0.370 \cdot 10^{-3}\left(\frac{\tau}{0.013}\right)^7\right]. \tag{15}$$

Take the case of a 2 cm diameter tube, for which the wall shear stress is $\tau_w = 5 \cdot 10^{-2}$ N/m^2. As $\tau(r)$ is known to be a linear function of r, one can calculate the shear rate distribution $\dot{\gamma}(r)$ by substitution in equation 15, and determine the axial velocity distribution v(r) by integration of $\dot{\gamma}$. By means of the integral $\int_0^R 2\pi r v(r)\, dr$, one obtains the flow Q which was calculated, for example, to be $Q = 1.026 \cdot 10^{-6}$ m^3/sec. For this known flow Q, the shear rate on the wall according to the Hagen-Poiseuille law would be given by: $\dot{\gamma}_w = 4\pi Q/3R^3$. In figure 16 is shown a comparison of the axial velocity profile v(r) and the shear rate distribution $\dot{\gamma}(r)$ for blood on the one hand, calculated from the above polynomial law, and a Newtonian fluid on the other, calculated according to the Hagen-Poiseuille law. The axial velocity v(r) was divided by the average velocity $\bar{v} = Q/\pi r^2$ for both blood and Newtonian fluid.

As can be seen from figure 16, the velocity profile of blood is considerably flattened when compared with that for a Newtonian fluid. The actual shear rate on the tube wall is approximately 27% above that calculated from the Hagen-Poiseuille law. However, the shear rate of blood and the Newtonian fluid are the same at the dimensionless radius

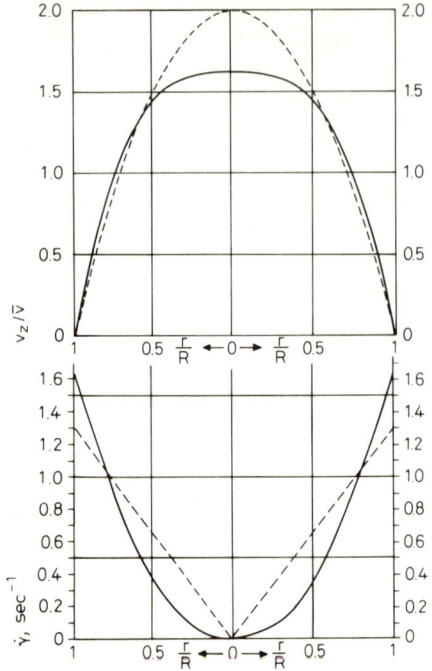

Fig. 16. Velocity (above) and shear rate distribution (below) in laminar pipe flow for Newtonian fluids (- - -) and normal human blood (—), with a haematocrit of 50%.

$r/R = \pi/4$. CHMIEL and SCHÜMMER [1971] have shown that this phenomenon applies accurately to all viscoelastic fluids.

For measured values of the flow Q and the relevant pressure loss Δp, sets of corresponding values of $\dot\gamma$ and τ at the radial distance $\pi R/4$ can be derived as follows:

$$\bar{\dot\gamma} = Q/R^3 \qquad (16)$$

and

$$\tau = \frac{\pi}{4} \cdot \frac{\Delta p}{\Delta L} \cdot \frac{R}{2} \qquad (17)$$

from which the flow curve of the fluid may be obtained.

It is presupposed in this case, as also for the validity of the velocity profiles illustrated in figure 16, that the flow, in addition to being completely formed and parallel, is also homogeneous, that is to say, the erythrocyte concentration is constant over the entire pipe cross-section.

This requirement, however, is not fulfilled for blood under all circumstances, as has been shown by CHISHOLM and GAINER [1971], GOLDSMITH [1971] and VOGTMANN et al. [1967]. FAHRAEUS and LINDQUIST had already pointed out, in 1931, that blood viscosity determined in narrow capillaries decreases with the tube diameter. This phenomenon is explained by the presence of a 'cell-free' blood plasma layer on the pipe wall. In blood rheology, this effect is named after its discoverers as the Fahraeus-Lindquist effect.

The phenomenon of particle migration from the wall has been known in suspension rheology for a long time [JEFFREY, 1932]. Although the existence of demixing in the proximity of the wall is agreed by FAHRAEUS [1921], SCOTT-BLAIR [1958], SCHULTZ-GRUNOW [1958], SEGRÉ and SILBERBERG [1961] and SAND [1963], there is a difference of opinion as to its cause. It was experimentally shown that a concentration thinning effect occurs in the proximity of the wall with polymer solutions and suspensions in all types of shear flow, that is to say, not only in flow along tubes, but also, for example, in the cone-plate or Couette systems. The smaller the ratio of the fluid gap to the particle diameter of the suspension, the more pronounced these effects become. For blood viscometry, therefore, it must be a general requirement that the thickness of the fluid layer of the respective measuring system (capillary, Couette, or cone-plate gap) should be as large as possible, in order to keep the ever-present separation effect to the minimum.

One may distinguish between roughly three diameter (D) ranges for blood flow in pipes, the limits of which, however, are still dependent on other parameters (erythrocyte diameter, haematocrit, Reynolds number, etc.) and are, therefore, incapable of being sharply defined:

(1) $D \geq 0.5$ mm, in which range blood can be regarded as a homogeneous fluid in the pipe, independent of the Reynolds number. The construction of velocity profiles from the flow curves (as in fig. 16) is possible for laminar flows.

(2) 0.5 mm $> D > 30$ μm: a demixing occurs on the wall which decreases with increasing Reynolds number, diameter and haematocrit. If one knows the thickness of the pure plasma zone, the velocity profile can be constructed, as the variable shear stress distribution is already known from reasons of balance. The plasma viscosity is used for the demixed zone, whereas the rheological curve of blood, for the relevant haematocrit, may be employed for the remainder of the range.

(3) $D \leq 30$ μm, in which range blood can no longer generally be regarded as a fluid, since the erythrocyte aggregates are of the same size as the tube diameter.

From this it may be concluded that a capillary viscometer can be

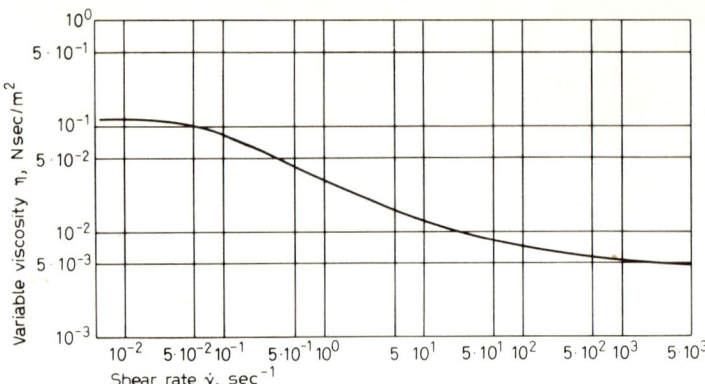

Fig. 17. Variable viscosity as function of shear rate for normal human blood with a haematocrit of 44% obtained from measurements using four different rheometers. After an initial region where η remains constant, the curve shows marked non-linearity and approaches a lower limit at high shear rates.

employed for blood viscosity measurement if the capillaries used have a diameter of $D \geqslant 0.5$ mm.

Figure 17 shows the variable viscosity as a function of the shear rate, for human blood with a haematocrit of 44%, which resulted from measurements with four different types of viscometers (wide and narrow gap Couette, cone plate and double capillary). The higher shear rate range was measured here with a recently developed double capillary viscometer which will be fully described in the next section. The pronounced non-linearity of the flow curve, as can be seen from figure 17, indicates that blood possesses elastic characteristics which considerably exceed those of plasma. Quantitative pronouncements, however, may only then be made when the normal stress functions F_1 and F_2 can be successfully measured technically. The recent work of THURSTON [1972] must be particularly mentioned here. The model on which his theory [THURSTON, 1960] is based is known in rheology as the 'Maxwell model':

$$\dot\gamma\eta_0 = t_0^* \dot\tau + \tau \tag{18}$$

where $\dot\tau$ is the rate of variation of shear stress with time and t_0^* is a relaxation time. This law approximates the flow behaviour of an actual viscoelastic fluid and is only valid in the range of very low stresses ($\eta \approx \eta_0$).

The apparatus used by THURSTON consists of a sinusoidally oscillating piston which displaces water in a chamber and thus displaces the same amount of blood in a second chamber which is separated from the first by an elastic membrane. In this way, a sinusoidally pulsating blood flow is produced in the system of parallel, rigid tubes of equal diameter and length which is placed above the second chamber. The necessary pressure is measured in the water-filled chamber. The 'apparent viscosity' of the blood, which is calculated from the measured data (i.e., pressure Δp, flow Q, frequency and the phase shift between pressure and flow), differs from the viscosity ascertained at stationary flow and equivalent shear rate.

Because of the oscillatory character of the flow, it is convenient to describe all quantities regarded in complex notation. According to THURSTON, the complex shear stress at the wall results from the complex expressed momentum balance

$$\Delta p^*/\pi R^2 - \tau_w^* 2\pi R = \rho \frac{\partial Q^*}{\partial t}. \tag{19}$$

The corresponding shear rate at the wall is

$$\dot{\gamma}_w^* = \frac{\tau_w^*}{\eta^*}. \tag{20}$$

For presentation, all quantities are shifted in phase, so that the imaginary part of the wall shear rate vanishes ($\dot{\gamma}_w^* = \dot{\gamma}_w$). The complex coefficient of the viscosity η^* consists of the viscous component η' and the elastic component η'':

$$\eta^* = \eta' - i\eta'' \tag{21}$$

or

$$\eta^* = |\eta^*| \exp(-i\theta). \tag{22}$$

θ means the phase angle of the complex viscosity which is related to the phase shift between flow and pressure. With the knowledge of one of the components of η^* and the corresponding viscosity η_0, one is in the position to calculate the relaxation time t_0^*. In order to describe oscillating flows in rigid and elastic vessels, at branch sites and in capillaries, however, it is absolutely necessary to know exactly the viscoelastic characteristics of blood at higher shear stress as well. While the flow curve of blood is already sufficiently well known from experimental work for the range of shear stress of interest here, new methods for the measurement of relaxation time must be developed to investigate the elastic characteristics.

IV. Blood Rheology in Clinical Pathology and Medicine

A. Viscometer for Clinical Application

The three blood rheological parameters which are important for the microcirculation are: (a) Plasma viscosity, (b) aggregation-tendency of erythrocytes, which rheologically can only be determined in the 'low shear' range, and (c) flexibility of erythrocytes, which is to be measured in the 'high shear' range.

The viscometers for the 'low shear' range have already been presented in section III. The suitability of the wide-gap Couette systems and sphere-sphere systems was pointed out. Because of the very low torque, however, very sensitive drive and measuring equipment is necessary. Therefore, in all cases where low shear rates are not absolutely necessary, devices with higher shear rate and accordingly lower sensitivity are to be preferred. For this purpose, a new capillary viscometer for clinical use was recently designed by CHMIEL [1975a, b]. The associated theory, which is based on a thesis of FREDRICKSON [1959], was developed by CHMIEL [1975a, b].

FREDRICKSON [1959] postulates that for two tubes of different length but equal flow and diameter, that part of the total pressure loss which is caused by entry and exit effects, is identical (fig. 18). Thus the total pressure difference between these two tubes represents the pure friction loss over the difference in length. The new capillary viscometer automatically eliminates end effects. It consists (fig. 19) of an electronically controlled synchronous motor (designated as 10 in the figure), which maintains its torque even at maximum load, and which drives a screw spindle (designated 9 here). A constant torque is thus applied to the nuts (7, 8) mounted on the spindle, which drives two identical pistons (6) with equal and constant velocities into two cylinders (4) displacing a fluid volume Q into each of the measuring chambers (2). Because of the incompressibility of the fluids – there is no gas in the system –, exactly the same volume Q of the substance to be measured is forced through each of the capillaries (1a and 1b). They are of equal diameter but of unequal length. The pressure difference Δp which builds up between the two measuring chambers corresponds to the pressure drop due to the fluid friction in the additional length L of the longer pipe, and is measured with a differential-pressure transducer (12). In this way, all end-effects are automatically eliminated; moreover, only the pressure difference need be measured and not the flow Q. Since the rotational speed of the spindle, its propulsion velocity and the diameters of the pistons and pipes are known, the flow Q is easily calculated.

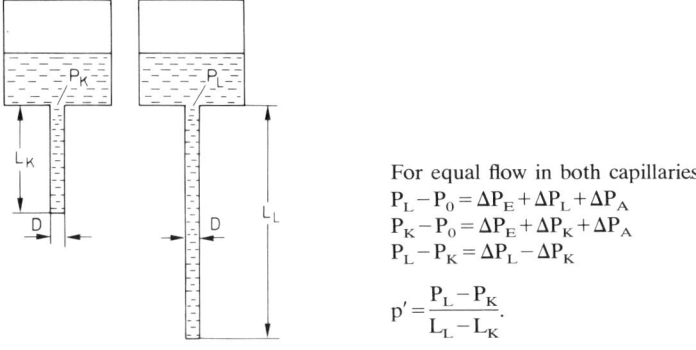

For equal flow in both capillaries:
$$P_L - P_0 = \Delta P_E + \Delta P_L + \Delta P_A$$
$$P_K - P_0 = \Delta P_E + \Delta P_K + \Delta P_A$$
$$P_L - P_K = \Delta P_L - \Delta P_K$$

$$p' = \frac{P_L - P_K}{L_L - L_K}.$$

Fig. 18. Flow through two tubes of different length having the same entry and exit conditions to flow. P_K and P_L are the pressures at the entry to capillaries, P_0 is the atmospheric pressure, ΔP_E and ΔP_A are pressure drops due to entry and exit effects, respectively, and ΔP_L and ΔP_K are the pressure drops due to friction in the long (L) and short (K) capillaries, respectively.

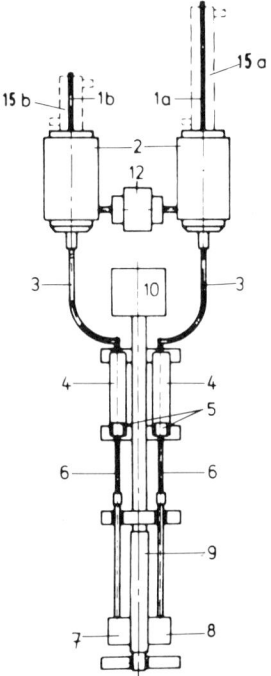

Fig. 19. Schematic diagram of the new double capillary viscometer according to CHMIEL. The difference in pressure, which builds up between the two measuring chambers (2) due to flow through capillaries of different lengths, is measured using a differential pressure transducer (12).

As one can see from figure 20, the pressure is not measured in the test-substance itself (9), but in the driving fluid (8) (e.g., water or a suitable oil), which is separated from it by a silicon-rubber membrane about 0.1 mm thick (4). The membrane is shaped like a test tube and is filled with the test fluid before being introduced into the measuring chamber (5). After the test, it is usually thrown away. The capillary tubes (1) are the only parts of the apparatus which require to be scrupulously cleaned. By means of a screw-cap (3), the membrane is hermetically sealed with a conical stopper (2) into which the measuring pipe is glued or soldered.

Figure 21 shows results obtained for blood plasma. The shear stress increases initially and reaches an end value which corresponds to the particular value of Q/R^3 (i.e., $\bar{\dot\gamma}$). Hence the viscosity, η ($=\tau/\dot\gamma$), is known. The accuracy obtained is very high. The entire test takes about 4 sec, and the test volume required is only about 1 ml of blood.

The technical advantage, of the pressure measurement device used in the double capillary viscometer, involves (i) the separation of the study fluid from the driving fluid, which enables one to simultaneously dispense with a large part of the tedious cleaning operations generally encountered in the use of the usual capillary viscometers, and (ii) the amenability to work with very small volumes of the test fluid. This has prompted the use of this device in rheometers designed for pulsatile flow measurements as well, described earlier. The result is the rheometer shown in figure 22. The lower part consists essentially of a Thurston type apparatus: a mini-shaker designed to produce a sinusoidally oscillating pressure pulse, which is transmitted via a spring-controlled oscillating piston, to the driving fluid (usually water) contained in the hermetically sealed chamber placed just on top. The pressure pulse is recorded in the transducer, fitted onto the wall of this pressure chamber, through the top of which is introduced (through a hermetically sealed screw joint) the small thimble-shaped silastic bag containing the test fluid (say blood). The silastic bag communicates with a capillary of precision bore, ending at the top in a small reservoir of the test fluid. The reservoir, the capillary and the silastic bag are all filled (avoiding enclosure of air bubbles) with the test fluid, which still requires only a small volume of it (≈ 2 ml).

The oscillating movement of the piston is monitored by a 'displacement monitor' which enables the calculation of the pulsating flow in the capillary under the influence of the monitored pressure pulses. A frequency response analyser serves to record the amplitudes of the fluid flow pulse in the capillary as measured through the displacement monitor and of the pressure pulse developed in the chamber, as also the phase shift between the two. An analysis of these data leads to a calculation of both

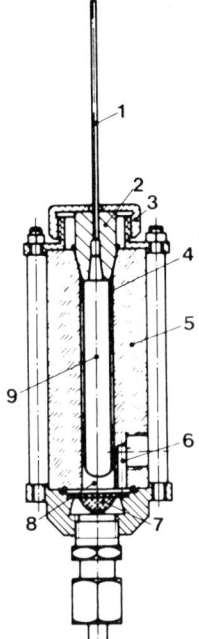

Fig. 20. Measuring chamber of the capillary viscometer. The substance under test (9) is contained in a silicone rubber thimble (4). Pressure is measured in the driving fluid (8) via the boring (6).

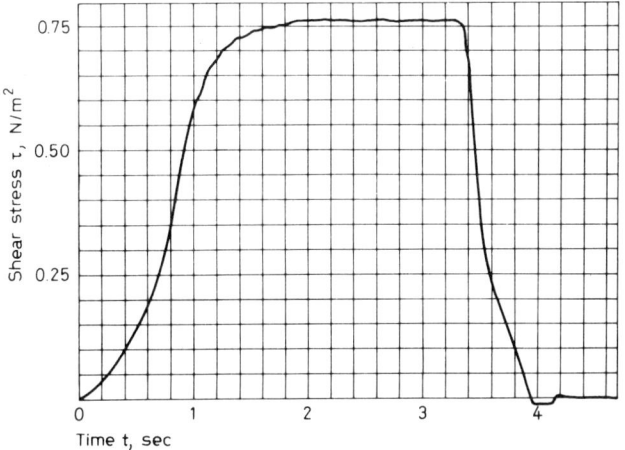

Fig. 21. Results obtained using the capillary viscometer for normal plasma. Time is shown on the abscissa, shear stress on the ordinate, shear rate $\dot{\gamma} = 430 \text{ sec}^{-1}$, temperature $t_0 = 23\,°C$.

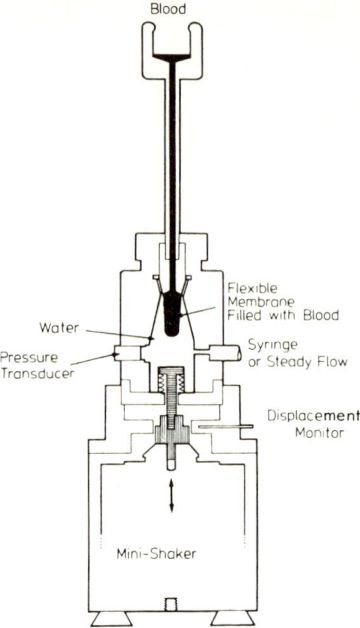

Fig. 22 New rheometer for the measurement of viscosity and viscoelasticity of blood and other biological fluids.

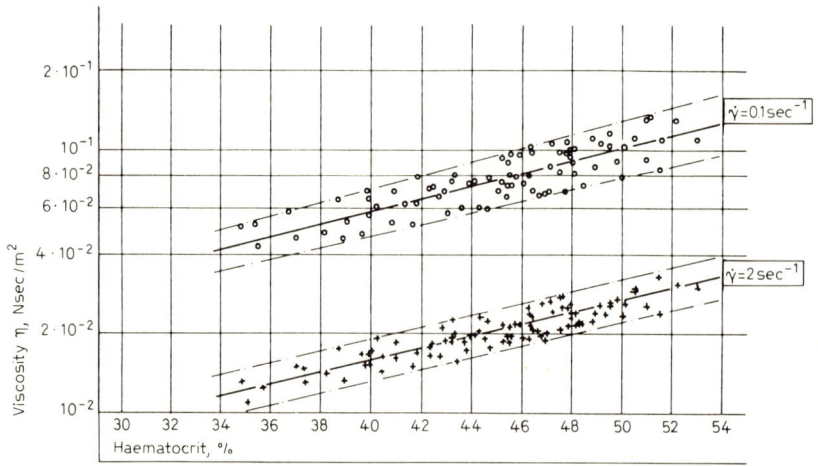

Fig. 23. Viscosity of normal human blood as function of the haematocrit for two different values of the shear rate, using a Couette viscometer with wide ($\dot{\gamma} = 0.1 \text{ sec}^{-1}$) and narrow ($\dot{\gamma} = 2 \text{ sec}^{-1}$) gaps.

B. Standard Values for Blood and Plasma

Venous blood of more than 150 healthy donors was investigated using the viscometers described above. The following procedure was found to be most convenient: venous blood and a sodium-citrate solution (3.8% Na-citrate in H_2O) in a volume ratio of 9:1 is drawn very slowly (danger of haemolysis!) into a 10-cm^3 plastic syringe. Two EDTA-coated plastic tubes are additionally filled with 8 cm^3 blood directly from the injection needle. The coating is necessary to keep haemolysis as low as possible, since the influence of haemolysis on blood and plasma viscosity is not known. For the measurement of the blood sedimentation rate (BSR), 4 cm^3 of blood are drawn into 1 cm^3 sodium-citrate solution.

Platelet-free plasma is obtained by centrifuging citrate-blood. Its viscosity is determined by a double capillary viscometer at 23 °C (fig. 20). The haematocrit (vol% of erythrocytes in blood) is determined by both haematocrit centrifuge and Coulter counter from one of the blood samples.

Apart from the haematocrit value, further information is needed for the characterisation of the blood, namely the number of leucocytes and erythrocytes, the erythrocyte volume, the haemoglobin content of whole blood and of erythrocytes, the density of whole blood and plasma, etc.

The contents of the EDTA-tubes are used for the determination of the flow-curve, for which measurements are made in the narrow Couette gap (LS), the wide Couette gap (LSW), the double capillary viscometer and the oscillating capillary rheometer. The BSR is determined simultaneously.

The mean plasma viscosity of 150 donors is found to be $1.73 \cdot 10^{-3}$ $Nsec/m^2$ (1.73 cP) at 23 °C and is independent of shear rate. The viscosity of whole blood is, however, a function of both the shear rate and the haematocrit value (see section III). Therefore, the viscosity of whole blood is depicted in figure 23 as a function of the haematocrit for a shear rate of $\dot{\gamma} = 0.1 \sec^{-1}$ in the wide Couette gap, and of $\dot{\gamma} = 2 \sec^{-1}$ in the narrow Couette gap. The choice of these particular values will be explained later. The relatively large scattering, of ±20%, at the low shear rate depends on a number of possible sources of error: Although evaluated by two different methods (micro-haematocrit centrifuge and Coulter counter), the determination of the haematocrit value is a possible error source. Other factors may be the influence of haemoglobin on the

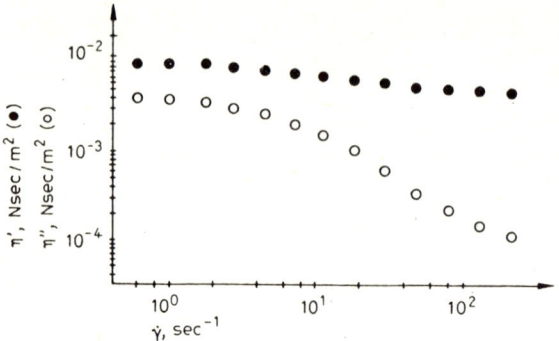

Fig. 24. Viscous (η') (●) and elastic (η'') (○) component of healthy blood as a function of shear rate; $f = 2$ Hz, $t_0 = 23\,°C$, haematocrit 45% [CHMIEL, 1978].

erythrocyte viscosity and an eventual oxygen uptake of the blood sample between extraction and measuring time (10–20 min).

During the search for further error sources an interesting phenomenon was detected. Heavy smokers – especially if they smoke immediately before blood extraction – have an increased blood viscosity (details will be given in context with the risk factors for heart infarction). This fact – which, of course, cannot be taken into account for the evaluation of the standard values – explains the increase of haematocrit-dependent scattering. The haematocrit determination by means of the Coulter counter showed an increased erythrocyte volume and hence an increased haematocrit value for smokers. Figure 23 shows that the average of the measured data can be adequately represented by a straight line on a logarithmic plot (continuous). The standard values of blood viscosity, when quoted in the following sections, will always refer to this average.

The viscoelasticity of whole blood was estimated by a frequency of 2 Hz, which is near to the *in vivo* pulsation in blood vessels. The temperature was always 23 °C. Figure 24 shows a typical plot of the viscous (η') and elastic (η'') components of healthy whole blood with a haematocrit of 45%.

C. Rheological Alterations of Blood Immediately after Heart Infarction

DINTENFASS [1971] already reported that blood viscosity is markedly increased after heart infarction. Since then a number of authors have contributed to this topic with partly contradictory results. ROSENBLATT *et al.* [1965], BEGG and HEARNS [1966] and ROZENBERG [1968], for instance,

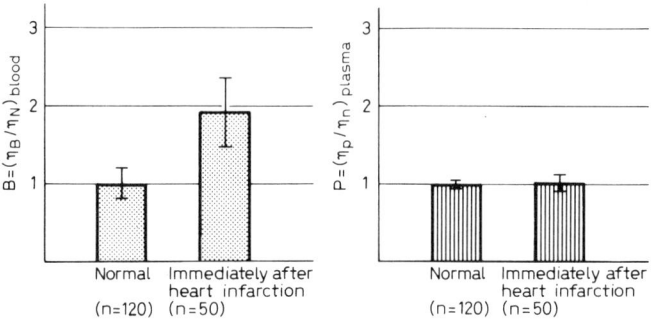

Fig. 25. Comparison of the blood and plasma viscosities shortly after heart infarction with the corresponding normal values. n refers to the number of donors. The vertical bars are the standard deviations.

state that blood viscosity is not or practically not increased after infarction, whereas LANGSJOEN and IMMON [1967], KALLIO *et al.* [1967] and BYDGEMAN and WELLS [1969] have measured a significant increase. The authors agreed, however, in the establishment of a markedly increased plasma viscosity, except for KAPRINEN [1970], who reported an increase of only 1.8%. Therefore, the author [CHMIEL, 1973] recently carried out an investigation to determine whether or not rheological changes appear after heart infarction and how the above-mentioned discrepancies could be explained. For this purpose, blood and plasma of more than 50 patients with recent heart infarction were investigated according to the methods described above. The time between actual infarction and measurement was 5–6 h, on an average. For the evaluation of the patient data, both the plasma viscosity η_p and the blood viscosity η_B (here at $\dot\gamma = 0.1\ \text{sec}^{-1}$) were related to their standard values:

$$P = \eta_p/\eta_N \tag{23}$$
$$B = \eta_B/\eta_N \tag{24}$$

Furthermore, the quotient $K = B/P$ is of interest.

Figure 25 demonstrates that the blood viscosity (at a shear rate of $\dot\gamma = 0.1\ \text{sec}^{-1}$) is about 50–150% above the standard value on the 1st day. The plasma viscosity, however, is practically normal. Figure 26 shows B, the ratio of measured blood viscosity to standard value at $\dot\gamma = 0.1\ \text{sec}^{-1}$ and P, the ratio of plasma viscosity to standard value, as functions of time. The 1st day is the day of the patient's arrival in hospital. In this case the blood viscosity increases further on the 2nd day and lies more than 200% above the standard value. Then the viscosity decreases steadily during a period of 10 days to a value of around 50% above the standard.

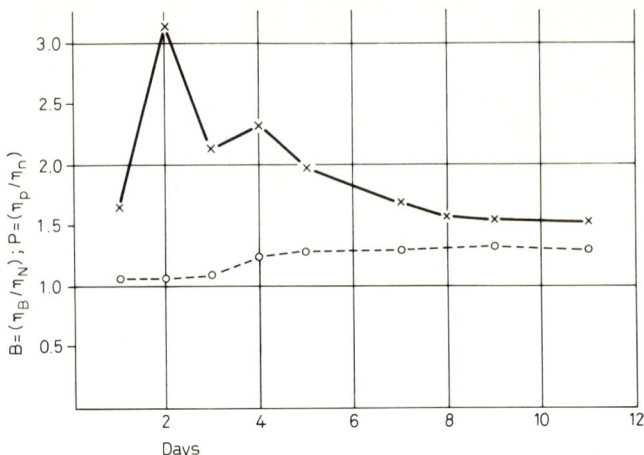

Fig. 26. The variation of B and P values following heart infarction as a function of time, measured at a shear rate of $\dot{\gamma} = 0.1\,\text{sec}^{-1}$. —— = Blood; - - - = plasma.

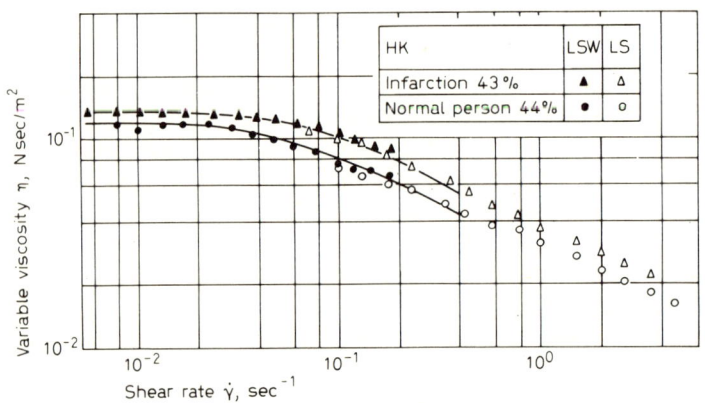

Fig. 27. Variable viscosity of a normal person and of an infarction patient. The increased viscosity arises due to aggregation of the erythrocytes. The effect is best observed at low shear rates. LSW indicates measurements made using a wide gap and LS, using a narrow gap.

The time variation of plasma viscosity was most surprising, since it remained nearly constant during the first 3 days. Then, however, it increased rapidly to a maximum value of nearly 30% above the standard. This type of time-viscosity course was particularly found in those cases where anticoagulants were not given in the first days. An analogous course was found by KALLIO *et al.* [1967].

At this stage the significance of K = B/P should be considered. The viscosity of a suspension depends linearly on the viscosity of the suspending fluid. Application of this fact to blood viscosity implies that if one wishes to ascertain whether an alteration of blood viscosity is due to an alteration of the suspended particles (mainly erythrocytes in this case) or due to a changed plasma viscosity, then the blood viscosity should be divided by the plasma viscosity. The K factor represents this normalisation. For clarity, K has not been used in figure 26. However, a calculation of K shows that (1) it decreases monotonically from the 2nd day on and (2) the increase in viscosity after infarction depends almost exclusively on a change in the tendency of the erythrocyte to aggregate.

The courses of plasma viscosity and BSR are in excellent agreement. The BSR after infarction is usually normal for the first 2 days and then increases rapidly. The increase of plasma viscosity and BSR, a few days after the infarction event, can be attributed to the rapid increase of fibrin and its products caused by heart muscle damage.

An explanation for the increased blood viscosity, while the plasma viscosity remains normal, can perhaps be found in a discussion of figure 27. Here the variable viscosity (as a function of the shear rate) of EDTA-blood of an infarction patient is compared with normal blood. Though the normal blood has a slightly higher haematocrit, its viscosity is significantly lower. It is remarkable that the zero viscosity of 'infarction' blood, apparently due to the increased tendency of the erythrocytes to aggregate, remains steadily higher with rising shear rate than the zero viscosity of normal blood. This increased tendency of the erythrocytes to aggregate may possibly be explained by an increased bonding capacity of hydrogen bonds. A higher load is required to decompose these aggregates. Recent results of our own investigations lead to the supposition that the number of available hydrogen bonds on the erythrocyte surface is influenced by certain plasma fractions. Figure 27 elucidates why a number of authors [ROSENBLATT et al., 1965; BEGG and HEARNS, 1965, and ROZENBERG, 1968] have found no changes in viscosity after infarction, whereas KALLIO et al. [1967], LANGSJOEN and IMMON [1967] and BYDGEMAN and WELLS [1969] noted distinct increases. If (for several shear rates) the ratio of the viscosity of infarction blood to the viscosity of normal blood (of the same haematocrit) is plotted against $\dot{\gamma}$, the resulting function shows a maximum value. The shear rate corresponding to this maximum, in fact, depends on the haematocrit and the degree of pathological viscosity alterations, and usually lies in the range of $\dot{\gamma} = 0.1 \text{ sec}^{-1}$. This is the reason why this special value of the shear rate was chosen for the calculation of the normal blood viscosity, as a function of haematocrit, in figure 23. As the shear rate is raised above this value, the viscosity differences between pathological and normal blood become smaller and smaller. Above a shear rate of $\dot{\gamma} = 200\ \text{sec}^{-1}$, no measurable difference could be found. If the literature is again screened from this point of view, it is noted that the authors who found no increase in viscosity after infarction, conducted their measurements in cone-plate rheometers at a constant shear rate of $\dot{\gamma} = 230\ \text{sec}^{-1}$. At this value, however, as shown above, the change is not measurable. On the other hand, the authors who found an increase, measured at a shear rate of $\dot{\gamma} = 23\ \text{sec}^{-1}$. These results, then, lead to the following conclusions: After infarction the blood viscosity is appreciably increased only in the region of low shear rates. The deviation from the normal value is greatest for a shear rate of about $0.1\ \text{sec}^{-1}$. The plasma viscosity remains normal during the first 4 days.

The increased blood viscosity has some possible haemodynamic consequences for the circulation. If one imagines the circulatory system to be one of rigid tubes and the heart a displacement pump, then for an imaginary experiment in which a 100% increase in effective viscosity occurs, this would result in a pressure duplication at the aortic valve. This concept is a gross simplification, since the arterial system is neither rigid, nor does the ventricle function as an ideal displacement pump and, furthermore, the complex physiological control system has been ignored. However, within the limitations of this control system it may be assumed that the natural circulation reacts to an increased viscosity in a threefold manner: (1) vasodilation (especially of the arterioles); (2) increased pressure in the greater arteries; (3) reduced heart minute volume. Because of the increased tendency of the erythrocytes to aggregate, the influence on microcirculation can become even more serious. When these aggregates reach the size of the capillary diameters, the capillaries can be occluded, causing a severe disturbance of the microcirculation.

D. The Influences of Medications on Blood Viscosity

For the reasons mentioned above, it is desirable to reduce the blood viscosity after infarction. EHRLY [1970] showed that the positive influence of streptokinase mainly depends on its considerable ability to reduce viscosity. Our own measurements confirmed these findings. According to EHRLY [1973], the snake venom Arvin has similar effects. It reduces the tendency for aggregation and increases the flexibility of the erythrocytes, thus causing a decreased blood viscosity. Our investigations have shown that most of the aggregation-restraining drugs have a more or less viscosity-reducing effect. The results obtained using heparin are still too poor to be able to make reliable conclusions.

E. Blood Viscosity – Related Risk Factors of Heart Infarction

In medicine, a number of factors are considered as risk factors for heart infarction. According to KANNEL et al. [1965] and VEDIN [1972], these factors include (among others): diabetes mellitus, cigarette abusus, hypertonicity, hypercholesterolaemia, adiposis and physical inactivity. A number of these factors were investigated in relation to changes in blood viscosity.

Investigations of blood viscosity in relation to diabetes are relatively scarce [SKOVBORG et al., 1966; DITZEL and SKOVBORG, 1966]. SKOVBORG et al. [1966] report viscosity measurement of 15 diabetic patients. They measured increases of viscosity of 10 and 18% at shear rates of 230 and 1.15 sec^{-1}, respectively. Plasma and serum viscosity data are not given. In a recently published study [STÖRMER et al. 1973], the flow curves of blood, plasma and serum of 12 diabetic patients were measured at shear rates between $5 \cdot 10^{-3}$ and 5 sec^{-1}. The variable viscosity of blood shows a course similar to that of infarction patients and lies, on average, 50% above the standard values (at $\dot{\gamma} = 0.1 \text{ sec}^{-1}$, $B = 1.5$). The serum viscosity of the 12 patients was about 13% higher than normal. This indicated that the increase of plasma viscosity is not just due to an increase of fibrinogen. Electrophoretic measurements confirm these findings: the albumin fraction is markedly reduced, whereas α_2- and β-globulins are strongly increased. SKOVBORG et al. [1966] suppose that the increased tendency for aggregation could be attributed to a 'mantling' of the erythrocytes with α_2-globulin and fibrinogen. Thus, the erythrocyte aggregates are responsible for disturbances of the microcirculation of diabetic patients.

As already reported in section IV. B, the blood viscosity of heavy cigarette smokers is significantly increased. The reduced blood perfusion of the extremities during or immediately after smoking was hitherto supposed to result from vasoconstriction. The present results indicate, however, that this reduced perfusion could be explained by an increased blood viscosity as well, which again is equivalent to an increased tendency of erythrocytes to aggregate. Cigarette smoking is statistically correlated with an increased risk of infarction. It is also known that about 2 weeks after stopping cigarette smoking, the risk is already reduced to that of a non-smoker. Therefore, it was tested whether the blood viscosity of heavy cigarette smokers (more than 20 cigarettes/day) changes when smoking is stopped for 24 h. For this purpose the whole test programme described in section IV.B was carried out.

Figure 28 shows an example of the viscosity measurements for a 30 cigarettes/day smoker. The upper curve represents the results im-

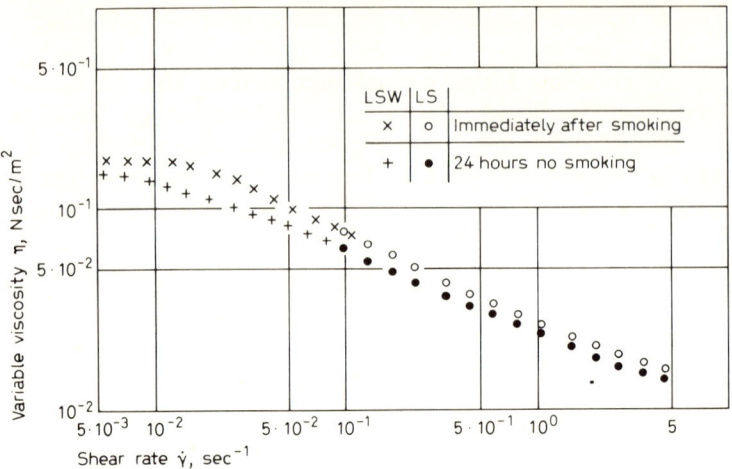

Fig. 28. Comparison of the variable viscosity of blood directly after smoking and 24 h later. LSW indicates measurements made using a wide gap and LS, using a narrow gap.

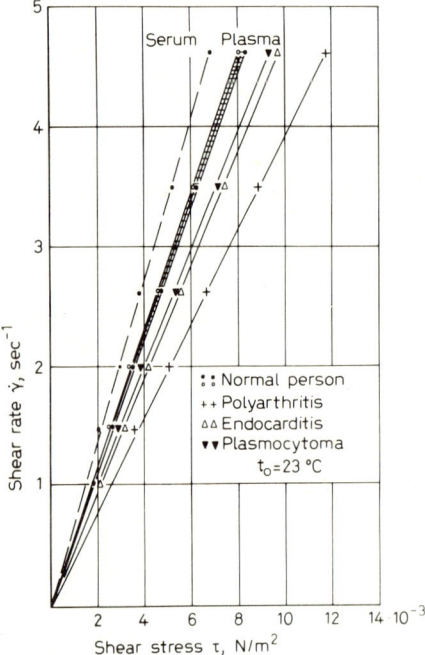

Fig. 29. Flow curve of healthy serum and plasma, and three pathological plasma samples.

mediately after smoking, the lower curve after a 24-hour interval of non-smoking. A significant lowering of viscosity is to be noted within this 24-hour period. Differences in plasma viscosity between smokers and non-smokers could not be found. These experimental results give a possible explanation for the question why heavy cigarette smokers have higher infarction risks than non-smokers, and why the risks fall so rapidly after smoking is discontinued.

At present, viscosity measurements of patients having hypercholesterolaemia, increased triglycerides and adiposis are being carried out. The results, however, are as yet insufficient to allow a statistical analysis.

F. Rheumatic Diseases

Measurements of plasma viscosity in rheumatic diseases have been reported already by HARKNESS [1971], BLADES *et al.* [1966] and DINTENFASS [1971]. In all cases an increased plasma viscosity was found. HARKNESS [1971] subdivides the patients with rheumatic arthritis into three groups: (1) active-chronic cases; (2) subacute cases; (3) acute cases. The average increase of plasma viscosity was 11% for the first, 30% for the second and 60% for the third group (the latter with a large standard deviation). BLADES *et al.* [1966] concluded that this increase is an index for the gravity of the disease. In a recent study [CHMIEL, 1973], the viscosity of blood and serum from 10 patients suffering from rheumatic diseases was measured. The plasma flow curve of a primary-chronic polyarthritis patient is depicted (among others) in figure 29. The plasma viscosity here is more than 45% higher than the standard value. The serum viscosity is markedly increased as well. This implies that the increase of plasma viscosity depends not only on increased fibrinogen. The electrophoretic measurements showed a decrease in albumin content and a strong increase in the α_1-, α_2- and β-globulins. For all 10 cases, the blood viscosity was above the standard value as well. The calculation of $K = B/P$ proved that the increased blood viscosity depends only on the increased plasma viscosity ($K \approx 1$).

G. Plasmocytoma

The blood of 6 patients with bone tumours (plasmocytoma) was studied in a further series of measurements. The whole programme described in section IV. B. was repeated. Figure 29 shows an example of a plasma flow curve. The calculated plasma viscosity lies clearly above the normal scattering range. In contrast to the rheumatic diseases, the plasma viscosity is relatively more increased than the serum viscosity. Hence it may be concluded that the plasma has a higher fibrinogen content. The electrophoretic measurements showed an unusually high portion of γ-globulins. The blood viscosity was increased in all 6 cases, and this was strictly correlated with an increase of the plasma viscosity ($K \approx 1$).

H. Von Willebrand-Jürgens Syndrome

As already reported elsewhere [ANGELKORT *et al.*, 1972], the measurement of the thrombocyteadhesivity after Hellem and the determination of plasma viscosity have considerably improved the diagnosis of the von Willebrand-Jürgens syndrome. The results show that this type of coagulation disturbance differs clearly rheologically from the uraemic one. In the case of the von Willebrand-Jürgens syndrome, the plasma viscosity is reduced about 10% on average, while it remains normal in uraemia. The reduction of the blood viscosity in cases of the von Willebrand-Jürgens syndrome is exclusively correlated with a reduction of the plasma viscosity ($K = 1$).

I. Anaemias

DINTENFASS [1971] and MURPHY [1967] have reported on rheological measurements made on blood from patients having different kinds of anaemias, such as sickle-cell anaemia, spherocytosis and Heinz-body anaemia. In all these diseases, the ratio of surface area to volume of the erythrocytes is increased, leading to a viscosity increase which rheologically differs clearly from the viscosity increase due to the increased aggregation of the erythrocytes during heart infarction and diabetes mellitus.

The anaemias mentioned above show a significant increase of viscosity at high shear rates and thus can be easily distinguished from those diseases which are accompanied by an increased tendency for erythrocyte aggregation.

The methods described for measuring the flow curve of blood, at high shear rates (double capillary viscometer) as well as at extremely low shear rates (wide Couette gap and sphere-sphere system), thus assist in the diagnosis of the anaemias mentioned. These methods have advantages over the rheological methods used hitherto. The advantage most often applied in testing is the filterability of erythrocytes using a microfilter.

J. Blood Sedimentation Rate and Plasma Viscosity

FAHRAEUS [1921] found a strong correlation between the degree of globulin concentration in plasma and the BSR. Since then, many experiments have been carried out in order to find a correlation between BSR and plasma viscosity as well. Our measurements in this context lead to the following conclusions:

For some diseases, as for instance heart infarction and plasmocytosis, there is a certain correlation between plasma viscosity and BSR; however, a general relation between these two parameters cannot be established. The very low divergence of plasma viscosity values for healthy subjects, the high reproducibility and the rapidity and simplicity of the plasma viscosity measurements are only some of its advantages in comparison with BSR measurements. Further advantages are discussed in the publication of HARKNESS [1971] cited above. He demonstrated that, in contrast to the plasma viscosity measurements, the BSR method cannot discriminate between the three mentioned groups of rheumatic diseases, the active-chronic, the subacute and the acute group. 33.5% of the rheumatic patients studied had a normal BSR, whereas only 6.7% had a normal plasma viscosity (the latter only in the active-chronic group). Similar results were found for tuberculosis. Thus it can be stated that the measurement of plasma viscosity is more informative and reliable than the BSR method.

K. The Viscoelasticity of Blood for Patients with Peripheral Vascular Diseases and for Heavy Smokers

The meaning of the viscoelasticity of blood for the microcirculation has been described by CHMIEL [1976]. One has to expect that an enhanced aggregation of erythrocytes will raise much more the elastic component than the viscous one. Indeed, a comparison of viscoelasticity of whole blood from healthy non-smokers, smokers and patients with peripheral vascular disease (table I) shows on the one hand that there is a significant increase in both the η' and η'' values of smokers and patients with peripheral vascular disease compared to the healthy donors; on the other hand that the elastic components are relatively more increased than the viscous one.

Figure 30 shows the plots of both components as a function of the shear rate for a healthy donor and a patient with peripheral vascular disease, both with a haematocrit of 45%. The plot of the elastic components exhibits a more striking increase at lower shear rates for patients

Table I. The viscous (η') and elastic (η'') components of viscoelasticity of whole blood from healthy donors, smokers and patients with peripheral vascular disease[1]

	n	$\eta' \cdot 10^{+3}$, Nsec/m^2 $\bar{x} \pm s$	p	$\eta'' \cdot 10^{+3}$, Nsec/m^2 $\bar{x} \pm s$	p
Healthy	64	7.4 ± 1.2		1.51 ± 0.47	
			<0.01		<0.01
Smoker	40	7.99 ± 1.01		1.77 ± 90.36	
			<0.0005		<0.0005
Peripheral vascular disease	13	8.86 ± 1.17		2.34 ± 0.61	

[1] $f = 2$ Hz; $t_0 = 23 \,°C$; $\dot{\gamma} = 10 \, sec^{-1}$.

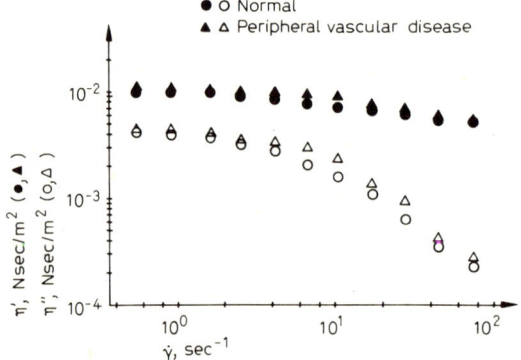

Fig. 30. Viscous (●, ▲) and elastic (○, △) components of blood as a function of shear rate for healthy donors (●, ○) and patients with peripheral vascular disease (▲, △); $f = 2$ Hz, $t_0 = 23 \,°C$ [CHMIEL, 1978].

with peripheral vascular disease compared to the plot of the healthy donor, thus indicating the increased aggregation of erythrocytes.

V. Conclusion

Over a wide range of shear rates the viscosity of human blood can be described mathematically by a polynomial of 7-th degree. This allows the calculation of the velocity distribution in different flow geometries, for example: flow in a tube, in a cylindrical gap, and in a right-angled channel. The pulsatile flow in geometries with nonparallel boundaries, which occurs everywhere in the natural vessels, will be influenced by the viscoelastic behaviour of the blood. For the range of low shear rates a new oscillating capillary rheometer allows the estimation of the elastic and viscous components of the viscosity of blood.

It has been shown that there is an increased tendency for erythrocytes to aggregate during heart infarction. These aggregation effects can only be measured in the range of very low shear rates and require a low-shear viscometer which minimises the surface effects at the air/blood interface simultaneously.

On the other hand, a change in the flexibility of the erythrocytes, which occurs in the case of several anaemias (for example spherocytosis), can be ascertained only at high shear rates. For the low shear region the sphere-sphere viscometer and the wide-gap Couette system should be used.

The new capillary viscometer, which was introduced in this chapter, is suitable for the measurement of both blood and plasma viscosity at high shear rates. This viscometer can possibly be used as a substitute for the BSR method, since HARKNESS found a correlation between plasma viscosity and BSR.

References

ANGELKORT, B.; CHMIEL, H.; HOLZHÜTTER, H. und WENZEL, E.: Funktionsfähigkeit, Aggregationstendenz und Adhäsivität von Thrombozyten im Vergleich zur Blutviskosität bei von-Willebrand Syndrom und Urämie. Medsche Welt, Stuttg. 23: 1808–1810 (1972).

BEGG, T. B. and HEARNS, J. B.: Components in blood viscosity, the relative contribution of hematocrit, plasma fibrinogen and other proteins. Clin. Sci. 31: 87 (1966).

BLADES, A. N.; COYER, A. B., and FLAVVEL, H. C. G.: Plasma viscosity with particular reference to its estimation in cases of rheumatoid-type arthritis. Ann. phys. Med. 6: 214–219 (1966).

BOSS, N.; CHMIEL, H.; KACHEL, K. und RUHENSTROTH-BAUER, G.: Erythrozytenaggregation bei Nichtrauchern, Rauchern und Herzinfarktpatienten. Blut 27: 191–195 (1973).

BRACHEL, H. V. und SCHÜMMER, P.: Möglichkeiten des Kugel-Kugel-Rheometers. Rheol. Acta 14: 219–224 (1975).

BUGLIARELLO, G. and SEVILLA, J.: Velocity distribution and other characteristics of steady and pulsatile flow in fine glass tubes. Biorheology 2: 85–107 (1970).

BYDGEMAN, S. and WELLS, R.: Studies on platelets' adhesiveness, blood viscosity and the microcirculation in patients with thrombotic disease. J. atheroscler. Res. 10: 33–39 (1969).

CHARM, S. E.: Static method for determining blood yield stress. Nature, Lond. 216: 1121–1123 (1967).

CHIEN, S.; USAMI, S.; DELLENBACK, R. J., and GREGERSEN, M. I.: Blood viscosity: influence of erythrocyte aggregation. Science 18: 829–831 (1967a).

CHIEN, S.; USAMI, S.; DELLENBACK, R. J., and GREGERSEN, M. I.: Blood viscosity: influence of erythrocyte deformation. Science 18: 827–829 (1967b).

CHIEN, S.; USAMI, S.; DELLENBACK, R. J., and GREGERSEN, M. I.: Shear-dependent deformation of erythrocytes in rheology of human blood. Am. J. Physiol. *219:* 136–142 (1970).

CHISHOLM, G. M. and GAINER, J. L.: Tube viscosimetry of blood: effect of wall material. J. appl. Physiol. *2:* 313–317 (1971).

CHMIEL, H.: Zur Blutrheologie in Medizin und Technik; Habilitationsschrift Aachen (1973).

CHMIEL, H.: New experimental results in hemorheology. Biorheology *11:* 87–96 (1974a).

CHMIEL, H.: Messung rheologisch komplexer Eigenschaften an verdünnten Polymerlösungen. Colloid Polym. Sci. *252:* 886–893 (1974b).

CHMIEL, H.: A new capillary viscometer for the clinical use. Biorheology *12:* 301–307 (1975a).

CHMIEL, H.: Ein neues Rohrrheometer zur Realisierung hoher Schergeschwindigkeiten. Rheol. Acta *14:* 246–251 (1975b).

CHMIEL, H.: Bemerkung zur Veröffentlichung Messung der Fliesskurve... Schergeschwindigkeiten. Rheol. Acta *14:* 666 (1975c).

CHMIEL, H.; EFFERT, S. und MATHEY, D.: Viskositätsuntersuchungen an gesundem und pathologischem Blut im Bereich extrem niedriger Beanspruchung. Z. exp. Chir. *6:* 237–247 (1973).

CHMIEL, H. und SCHÜMMER, P.: Eine neue Methode zur Auswertung von Rohrrheometerdaten. Chem.-Ing. Techn. *43:* 1257–1259 (1971).

CHMIEL, H. und SCHÜMMER, P.: Messung der Fliesskurve im Bereich extrem niedriger Schergeschwindigkeiten. Rheol. Acta *11:* 1–3 (1972).

CHMIEL, H., und STÖRMER, B.: Zur Rheologie des Blutes. Biomed. Techn. *5:* 174–180 (1972).

CHMIEL, H.; THURSTON, G. B. und EFFERT, S.: Die klinische Bedeutung der Viscoelastizität des Blutes. Verh. dt. Ges. Kreislforsch. *42:* 376 (1976).

CHMIEL, H.; ANADERE, I.; HESS, H., and THURSTON, G. B.: Clinical blood rheology. 3rd Int. Congr. Biorheology, La Jolla 1978. Biorheology (to be published, 1979).

COKELET, G. R.; MARRILL, E. W.; GILLILAND, E. R., and SHIN, H.: The rheology of human blood-measurement near and at zero shear rate. Trans. Soc. Rheol. *7:* 303–317 (1963).

COPLEY, A. L.: Non-Newtonian behaviour of surface layers of human plasma protein systems and a new concept of the initiation of thrombosis. Biorheology *2:* 79–84 (1971).

COPLEY, A. L. and KING, R. G.: Viscous resistance of thromboid (thrombus-like) surface layers in systems of plasma proteins including fibrinogen. Thromb. Res. *1:* 1–18 (1972).

COWIN, S. C.: On the polar fluid as a model for blood flow in tubes. Biorheology *2:* 67–82 (1972).

DINTENFASS, L.: Blood microrheology. Viscosity factors in blood flow, ischaemia and thrombosis (Butterworths, London 1971).

DITZEL, J. and SKOVBORG, F.: Hemorheological investigations in relation to diabetes mellitus and its angiopathy. Proc. 1st Int. Conf. Haemorheology, Reykjavik 1966 (Pergamon Press, Oxford 1968).

EHRLY, A. M.: Zur Wirkung von Streptokinase bei Herzinfarkt. Rheologische Untersuchungen. Die gelben Hefte *19:* 973–978 (1970).

EHRLY, A. M.: Influence of Arvin on the flow properties of human blood. Biorheology *3:* 453–456 (1973).

FAHRAEUS, R.: The suspension-stability of the blood. Acta med. scand. *55:* 3-228 (1921).

FAHRAEUS, R. and LINDQUIST, T.: The viscosity of blood in narrow capillary. Am. J. Physiol. *96:* 562 (1931).

FREDRICKSON, A. G.: Flow of non-Newtonian fluids in annuli; thesis, Univ. of Wisconsin (1959).
GOLDSMITH, H. L.: Deformation of human red cells in tube flow. Biorheology 4: 235–242 (1971).
GOLDSTONE, J.; SCHMIDT-SCHÖNBEIN, H., and WELLS, R. E.: The rheology of red blood cell aggregates Microvasc. Res. 2: 273–286 (1970).
HARKNESS, J.: The viscosity of human blood plasma; its measurements in health and disease. Biorheology 3–4: 171–193 (1971).
HERSHEY, D. and CHO, S. J.: Blood flow in rigid tubes: thickness and slip velocity of plasma film at the wall. J. appl. Physiol. 21: 27–32 (1966).
JEFFREY, G. B.: The motion of ellipsoidal particles immersed in a viscous fluid. Proc. R. Soc. A 102: 161 (1932).
KALLIO, V.; SALMIVALLI, M., and BRUMMER, P.: Blood viscosity changes in patients hospitalized because of acute chest pain. Cardiologia 50: 323–329 (1967).
KANNEL, W. B.; WIDMER, L. K. und DAWBER, T. R.: Gefährdung durch coronare Herzkrankheit. Schweiz. med. Wschr. 95: 18–24 (1965).
KAPRINEN, K.: The electrophoretic mobility of red cells and platelets and the plasma viscosity in coronary heart disease. Acta med. scand. suppl. 506 (1970).
LANGSJOEN, P. H. and IMMON, T. W.: Rheologic changes in myocardial infarction. Am. Heart J. 73: 430–431 (1967).
MURPHY, J. R.: The influence of pH and temperature on some physical properties of normal erythrocytes and erythrocytes from patients with hereditary spherocytosis. J. Lab. clin. Med. 69: 756–761 (1967).
OKA, S.: The principles of rheometry; in EIRICH, Rheology, vol. 3 (Academic Press, New York 1960).
ROSENBLATT, G.; STOKES, J., and BASSET, D. R.: Whole blood viscosity, hematocrit and serum lipid levels in normal subjects and patients with coronary heart disease. J. Lab. clin. Med. 2: 202–211 (1965).
ROZENBERG, M. C.: Blood viscosity and thrombosis *in vitro* in patients with previous myocardial infarction. Angiology 19: 527 (1968).
SAND, J. P.: Entmischungseffekte in nicht-sedimentierenden Suspensionen runder Teilchen; Diss. Aachen (1963).
SCHMIDT-SCHÖNBEIN, H. and WELLS, R. E.: Quantification of the dynamics of red cell aggregation. Biblthca anat., Nr. 10, pp. 45–51 (Karger, Basel 1969).
SCHMIDT-SCHÖNBEIN, H. and WELLS, R. E.: Rheological properties of human erythrocytes and their influence upon the 'anomalous' viscosity of blood. Ergebn. Physiol. 63: 146–219 (1971).
SCHMIDT-SCHÖNBEIN, H.; WELLS, R. E., and SCHILDKRAUT, R.: Microscopy and viscometry of blood flowing under uniform shear rate. J. appl. Physiol. 5: 674–678 (1969).
SCHULTZ-GRUNOW: Entmischung makromolekularer Lösungen in Scherströmungen. Rheol Acta 1: 298 (1958).
SCOTT-BLAIR G. W.: The importance of the sigma phenomenon in the study of the flow of blood. Rheol. Acta 2–3: 123–126 (1958).
SEGRÉ, G. and SILBERBERG, A.: Radial particle displacement in Poiseuille flow of suspensions. Nature, Lond. 189: 209 (1961).
SKALAK, R.; CHER, P. H., and CHIEN, S.: Effects of hematocrit and rouleaux on apparent viscosity in capillaries. Biorheology 2: 67–82 (1972).
SKOVBORG, F.; NIELSEN, A. V.; SCHLICHTKRULL, J., and DITZEL, J.: Blood-viscosity in diabetic patients. Lancet i: 129–131 (1966).

Störmer, B.; Chmiel, H.; Potthoff, D.; Kremer, K. und Brüster, H.: Viskositätsuntersuchungen an gesundem und pathologischem Blut im Bereich extrem niedriger Beanspruchung. Z. exp. Chir. 6: 237–247 (1973).

Stoltz, J. F. et Larcon, A.: Determination du seuil de contrainte pour le sang. Biorheology 2: 129–135 (1970).

Thurston, G. B.: Theory of oscillation of a viscoelastic fluid in a circular tube. J. acoust. Soc. Am. 2: 210–213 (1960).

Thurston, G. B.: Measurement of the acoustic impedance of a viscoelastic fluid in a circular tube. J. acoust. Soc. Am. 2: 1091–1095 (1961).

Thurston, G. B.: Viscoelasticity of human blood. Biophys. J. 12: 1205 (1972).

Vedin, A.: Plötzlicher Herztod: Epidemiologie und Risikofaktoren. Medsche Klin. 67: 752–753 (1972).

Vogtmann, C.; Gerbstädt, H. und Rüth, P.: Strömungsverhalten und Scheinviskosität von heparinisiertem menschlichem Blut und Zitratblut in Glaskapillaren mittlerer Weite. Acta biol. med. germ. 18: 499–505 (1967).

Wells, R. E. and Merrill, E. W.: Shear rate dependence of viscosity of human blood and blood plasma. Rheol. Bull. 30: 79–84 (1961).

Prof. Dr.-Ing. H. Chmiel, Institut für Grenzflächen- und Bio-verfahrenstechnik, Eierstrasse 46, *D-7000 Stuttgart-S* (FRG)

Blood-Gas Interactions and Physiological Implications

W. J. YANG

Mechanical Engineering Department, University of Michigan,
Ann Arbor, Mich.

Contents

Abstract	46
List of Symbols	46
I. Introduction	48
II. Gas Exchange in the Lungs and Its Transport by the Blood	49
A. The Lungs and Gas Exchange	49
1. Composition of Atmospheric and Alveolar Airs	49
2. Diffusion of Oxygen and Carbon Dioxide through the Alveolar-Capillary Membrane	50
3. Rate of Gas Uptake in Lung Capillary Blood	52
B. The Carriage of Oxygen and Carbon Dioxide in the Blood	55
III. Gas Emboli in Human Body	57
A. Sources of Gas Emboli	57
1. Accidentally Introduced	57
2. Purposely Introduced	58
B. Expansion and Dissolution of Gas Emboli in Blood	59
1. Theory	59
a) The Case Where Mass Diffusion Controls	59
b) The Case Where Liquid Inertia Controls	65
2. Experiments	67
C. Effects of Foreign Agents on the Behavior of Gas Emboli	70
1. Plasma Substitutes	70
2. Anesthetics	71
IV. Gas Embolism Due to Extracorporeal Oxygenation	72
A. Artificial Heart-Lung Machines	72
B. Blood Trauma	73
1. Possible Factors Responsible for Hemolysis	73
2. Denaturation	74
C. Introduction of Gas Emboli into the Blood during Open-Heart Surgery	74
D. Hydrodynamic Basis of Hemolysis	75
1. Sphering of Erythrocytes	75

          2. Critical Membrane Yield Stress 75
          3. Wall and RBC Collisions 76
          4. Shock Waves and Liquid Jets Produced by Collapsing Emboli 76
     E. Gas Embolism Syndromes, Prevention and Treatment 77
          1. Types of Air Embolism 77
          2. Treatment and Prophylactic Measures 79
V. Gas Embolism Due to Sudden Decompression 79
     A. Bubble Formation Due to Sudden Decompression 80
          1. Nucleation 80
          2. Stable Nuclei 81
          3. Criteria of Cavitation 81
          4. Factors Affecting Bubble Growth or Shrinkage 82
     B. Symptoms of Decompression Sickness 82
     C. Prevention and Treatment of Decompression Sickness 83
VI. Tissue-Capillary Gas Exchange 84
     A. Exchanges of Oxygen and Carbon Dioxide in the Tissues 85
          1. Exchange of Oxygen 85
          2. Exchange of Carbon Dioxide 86
     B. Dissolution of Gas Emboli in the Tissues 86
          1. *In vivo* Tests.. 87
          2. Theory 91
               a) Tissue Creep Controlling 92
               b) Mass Transfer Controlling 93
               c) Both Tissue Creep and Mass Transfer Controlling. 96
          3. Comparison of Theory with *in vivo* Tests 97
References 97

Abstract

Blood-gas interaction in the lungs and tissues is treated by classifying the gas interchanges into three main stages, namely, external exchange, transport and internal exchange. Sources of gas embolism introduced during extracorporeal oxygenation, through sudden decompression and for in vivo tonometry are presented together with the theory and experiments on their growth and dissolution under various environmental conditions and the effects of foreign agents such as plasma substitutes and anesthetics on gas embolism. Hydrodynamic aspects of hemolysis, treatment and prophylactic measures of gas embolism syndromes, and prevention and treatment of decompression sickness are also discussed.

List of Symbols

A	diffusion area, cm^2		(ii) and (iii) in section VI, $mm\,Hg\,sec^{-1}$
A_c	capillary cross-sectional area, cm^2		
$\bar{A}$	surface area of a capillary tube, $=\bar{p}x$, cm^2	B^*	$B/(p_g-p_\infty)$, sec^{-1}
a	radius of a sphered red corpuscle, cm	C	concentration of dissolved gas in liquid, ml gas/ml liquid or g/ml; C_s, saturated value; C_∞, value at a large
B	q/α_t for case (i) or $\dot{Q}\Delta C/\alpha_b$ for cases		

C_ρ^*	C_s/ρ_g		distance from a gas bubble or pocket
C_∞^*	C_∞/C_s		tension; P_∞, hydrostatic pressure in blood
D	diffusion coefficient of dissolved gas in blood, cm²/sec; D_M, in pulmonary membrane; D_w, in water	P_b	mean blood pressure, mm Hg
		P_{crit}	critical buckling pressure of biconcave RBC membrane, dyn/cm²
D_L^*	diffusion capacity of the lungs as defined by equation 3, ml/(sec · mm Hg)	P_{jet}	pressure produced by the impact of liquid jet on surface, dyn/cm²
		P_{max}	maximum liquid pressure, dyn/cm²
D_p	Krogh-type diffusion coefficient, $=\alpha D$, cm² mm Hg⁻¹ sec⁻¹	$\bar{P}$	perimeter of a capillary tube, cm
		$\dot{Q}$	rate of blood perfusion per unit tissue volume, sec⁻¹; Q_p, pulmonary; Q_t, in subcutaneous tissue
D^*	diffusing capacity $\alpha DA/x$, ml mm Hg⁻¹ sec⁻¹, defined in equation 1; D_M^*, in pulmonary membrane		
		q	rate of oxygen consumption per unit tissue volume, sec⁻¹
d	differential operator	R	bubble or pocket radius (mean radius for nonspherical one), cm; R_c, critical value for stable bubble; R_0, initial value
E	$\alpha_b k \dot{Q}_t/\alpha_t$, sec⁻¹		
F	FP_b = amplitude of blood pressure pulse, mm Hg		
f	fP_b = amount of step change in external pressure, mm Hg	$\bar{R}$	gas constant, mm Hg · ml/(g · °K); $\bar{R}_j$, of j-th component gas
g_c	conversion factor, = 980 cm sec⁻²	$\bar{R}_g$	universal gas constant, =1.987 cal g · mol⁻¹ °K⁻¹
H	α_t/ρ, ml/(g · mm Hg)		
I_i	function as defined by equation 54, cm⁻¹	$\dot{R}, \ddot{R}$	first- and second-time derivatives of R
K	k' times concentration of reduced hemoglobin, sec⁻¹	R^*	R/R_0
		r	radial distance from bubble center, cm
K^*	kR_0^2/D		
k	coefficient of end capillary saturation	S	parameter defined by equation 31
k'	reaction velocity constant for the association of O_2 and Hb to form HbO_2, sec⁻¹	T	temperature of blood, tissue-capillary system, or bubble, °C
		t	time, sec
k''	reaction velocity constant for the dissociation of O_2 from HbO_2, sec⁻¹	t_f	bubble lifetime (time required for complete collapse), sec
l	thickness of capillary wall, cm	t^*	Dt/R_0^2
M	molecular weight of gas inside bubble, g	U	liquid jet velocity, cm/sec, or dR/dt
		u	blood velocity in pulmonary capillaries, cm/sec
m	mass of gas inside bubble, g; m_0, initial value; m_j, of j-th component gas; m_{0j}, initial value of m_0		
		$\dot{V}_A$	alveolar ventilation rate per unit tissue volume, ml/(sec · ml)
N	rate of gas diffusion, ml/sec or g/sec	V_c	blood volume in a capillary tube, $=A_c x$, ml
n	polytropic exponent; $n=1$ for isothermal process	x	location in a capillary tube measured from venous blood entrance, cm
P	partial pressure of gas, mm Hg; P_A, of alveolar gas; P_c, in capillary blood; $P_{c0}(P_c)$, at inlet of venous blood; P_g, inside gas bubble or pocket; $P_{g0}(P_g)$, at initial state; $P_{gj}(P_g)$, for j-th gas; P_i, inside red corpuscle; P_0, outside red corpuscle; P_P, in plasma; P_R, in red corpuscle; P_t, gas	Greek letters	
		α	solubility coefficient, $=C/P$, ml gas/ml liquid, mm Hg; α_B, in blood; α_M, in pulmonary membrane; α_t, in tissue; α_w, in water
		β	coefficient of volume compressibility

	of blood, dyn/cm²	Superscripts	
γ	ρ*+σ*/R*	., ..	first- and second-time derivatives, respectively, or rate
ε_{rr}	strain in tissue-capillary system		
η	rheological constant of tissue-capillary system, dyn/cm²	Subscripts	
Θ	rate of gas uptake, ml/(min · mm Hg · ml of blood)	A	alveolar
		b	in bold
θ	angle in spherical coordinates, rad	c	cross section for A, capillary for P, conversion for g, or critical value for R
λ	rheological constant of tissue-capillary system, sec		
λ*	ratio of suspended to suspending phase viscosity	crit	critical value
		g	gas
μ	rheological constant of tissue-capillary system, sec, or blood viscosity, dyn · sec/cm²	i	integer; 1 for single-gas diffusion case; 2 for multi-gas diffusion case, or inside red corpuscle
ρ	blood density, g/ml	j	j-th component gas
π_g	density of gas inside bubble, g/ml or ml gas/ml liquid; ρ_{g0}, initial value of ρ_g; $\rho_{g\infty}$, ρ_g under the same conditions of pressure and temperature with a gas-liquid surface of zero curvature	jet	of liquid jet
		M	in pulmonary membrane
		max	maximum value
		P	in plasma
		R	in red corpuscle
ρ*	$\rho_{g\infty}/\rho_{g0}$	s	at saturate state
σ	coefficient of surface tension, dyn/cm	t	in tissue
σ*	$2\sigma/(R_0 P_{g0})$	w	in water
τ	shear stress in tissue-capillary system or in liquid, dyn/cm²; τ_{rr} and $\tau_{\theta\theta}$, normal components in the r and θ direction, respectively; τ_∞, $=\tau_{rr}(\infty, t)$ at a large distance from bubble; τ_{rr0}, initial value of τ_{rr}	0	initial value (at zero time), at inlet of venous blood, or outside red corpuscle
		∞	hydrostatic value in blood, or at a large distance from bubble (r → infinity)
τ_{jet}	shear stress in blood produced by jet, dyn/cm²		
ω	blood pressure pulse, sec⁻¹	Note: liquid refers to blood	

I. Introduction

In man as well as all other living organisms, uninterrupted exchanges of chemical substances are carried out between the cells which are composed of these materials and the external environment in which these cells live. These *interchanges take place in three main stages: external exchange, transport and internal exchange. The external exchange* refers to the exchange of materials including oxygen, nutriment, carbon dioxide and waste between the exterior and certain organs such as lungs, kidneys, digestive tract and others, which are situated at the external boundaries of the organism. *The transport* of these chemical substances between the organs where external exchanges are carried out and the cells which consume oxygen and nutriment to produce carbon dioxide and waste in metabolism is provided by certain circulating liquids, for example the blood stream in man and many other organisms. *The internal exchange* is the interchange of chemical substances between the blood stream and the metabolizing cells.

II. Gas Exchange in the Lungs and Its Transport by the Blood

A. The Lungs and Gas Exchange

The lungs act as a gas exchanger, which supplies oxygen to the tissues and organs of the body and discharges the carbon dioxide waste produced by cell metabolism. The main airways in the lungs are called *bronchi* which give rise to several million terminal passages through repeated branching. Each of these small ducts ends by opening into numerous blind, gas-filled pockets, called the *alveoli*. There are about 300 million alveoli in the two lungs. Their diameters range from 80 to 300 μm depending upon the degree of expansion of the chest. The total surface area of the alveolar walls is estimated to be 100 m^2, as compared to 2 m^2 for the surface area of the body.

The air in the alveoli is separated from the blood in the capillaries by the *alveolar-capillary* (or pulmonary) *membrane* which has a thickness of only 0.2–0.4 μm. As the blood traverses the capillary, oxygen diffuses from the alveolar air into the blood while carbon dioxide diffuses from the blood into the alveolar air.

1. Composition of Atmospheric and Alveolar Airs

One of the principal factors that determines *the rate at which a gas will diffuse through the pulmonary membrane* is the difference between the partial pressures of the gas on the two sides of the membrane. Table I

Table I. Partial pressures of respiratory gases in the atmosphere and in the alveoli

Gas	Atmospheric air		Alveolar air	
	partial pressure, mm Hg		partial pressure, mm Hg	
Nitrogen	597.0	78.62	569.0	74.9
Oxygen	159.0	20.84	104.0	13.6
Carbon dioxide	0.15	0.04	40.0	5.3
Water vapor	3.85	0.5	47.0	6.2

shows the relative pressures of nitrogen, oxygen, carbon dioxide, and water vapor in atmospheric and alveolar airs. As air is inspired, it is humidified immediately by the moisture on the linings of the respiratory pathways and alveoli. This results in an increase in the partial pressure of water vapor from 3.85 mm Hg in the atmosphere to 47 mm Hg in the alveolar spaces at normal body temperature. The total pressure of the gases (including water vapor) remains the same as the barometric pressure. However, since 47 mm Hg of that pressure is taken up by water vapor, the pressure of the other gases is the barometric pressure minus 47 mm Hg. *The actual partial pressure of one of the gases in the alveolar space* is then calculated by multiplying the fraction of the gas in dry gas by the barometric pressure minus 47 mm Hg, as given in table I.

2. Diffusion of Oxygen and Carbon Dioxide through the Alveolar-Capillary Membrane

Fresh air being drawn into the alveoli diffuses completely in the alveolar space within a fraction of a second. The molecules of oxygen in the gas mixture then diffuse through the alveolar-capillary membrane into the capillary blood.

Fick's first law of diffusion can be expressed as

$$N = DA \frac{\Delta C}{x} = D_p A \frac{\Delta P}{x} = D^* \Delta P, \tag{1}$$

where x is the thickness of the medium with concentration difference ΔC, while the other quantities are defined in the 'List of Symbols'. The first expression is most commonly used in physical sciences for diffusion in the same phase, while the second one is obtained through the application of Henry's law:

$$C = \alpha P.$$

The latter is particularly necessary when gases diffuse through different phases, most commonly observed in biological work. The so-called 'diffusing capacity' D^* in the last expression of equation 1 is equivalent to a conductance in terms of an electrical analogue. The magnitudes of D and α of the non-inert gases O_2, CO and CO_2 and the inert gas N_2 (does not combine chemically with the components of blood, but only dissolves physically in them) are listed in table II. The solubility coefficient in the pulmonary membrane can be approximated as that in water.

The total pressure of atmospheric air is normally 760 mm Hg and of this, 20.84% is due to oxygen. That means that PO_2 of air is 28.04% of 760 mm Hg. As air is breathed, it becomes saturated with water vapor at

Table II. Solubility and diffusion coefficients of gases in blood and water at STP

	Water at 20 °C		Blood	
	α_W ml gas/(ml·atm)	$D_W \times 10^5$ cm²/sec	α_B ml gas/(ml·atm)	$D \times 10^5$ cm²/sec
O_2	0.0314	1.98	0.023	
CO_2	0.872	1.77	0.49	
CO	0.0232		0.018	
N_2	0.0155	2.02	0.012	2.5

Table III. Partial pressures and dissolved gas concentrations in the lungs

	PO_2 mm Hg	Dissolved O_2, vol%	PCO_2 mm Hg	Dissolved CO_2, vol%	PN_2 mm Hg
Alveolus (alveolar air)	104		40		569
Venous blood	40	0.12	45	2.9	567
Arterial blood	100	0.30	40	2.6	569
Tissue cell	≅30		55		

body temperature. The PO_2 of moist inspired gas is therefore $0.2084 \times (760-47) = 149$ mm Hg. However, the alveolar gas contains only 14% of oxygen, corresponding to a PO_2 of $0.14 \times (760-47)$ mm Hg or about 100 mm Hg. Since the PO_2 of the capillary blood to be oxygenated is approximately 40 mm Hg, *there is the driving force of 60 mm Hg for the diffusion of oxygen molecules between gas and blood.* When the venous blood is made arterial, the PO_2 reaches about 96 mm Hg, indicating that the membrane exerts a negligible resistance to oxygen diffusion.

In addition to supplying the tissues with oxygen, the circulating blood also removes the waste products of cellular activity such as excess carbon dioxide. *The partial pressure of carbon dioxide* in alveolar gas is about 40 mm Hg, while the PCO_2 *of venous blood* entering the lung capillaries is about 46 mm Hg under resting conditions. The rate of diffusion of carbon dioxide molecules can be calculated by equation 1. The diffusion of carbon dioxide molecules across the alveolar-capillary membrane is so rapid that when the blood has completely traversed the lung capillaries and is made arterial, its PCO_2 is identical with the PCO_2 in the alveolar gas. This signifies that when the alveolar gas PCO_2 is changed, the PCO_2 of the arterial blood will change by an exactly similar amount. *The PCO_2 in the arterial blood is the fundamental factor regulating the rate and depth* *of breathing*. For example, if the PCO_2 falls below 40 mm Hg, breathing is inhibited temporarily until the PCO_2 in the body is restored to normal. On the contrary, when the PO_2 exceeds 40 mm Hg, breathing increases temporarily until normal PCO_2 is restored.

In the resting adult, about 250 ml of oxygen per minute is used by the metabolizing cells, while about 200 ml of carbon dioxide is produced in the same period of time. During muscular exercise, both the alveolar ventilation and the lung blood flow may increase linearly up to an oxygen consumption of 3,000–3,500 ml/min.

3. Rate of Gas Uptake in Lung Capillary Blood

Blood flows rapidly through the lungs. A single erythrocyte passes through the lung capillaries in about 0.75 sec at rest and about 0.3 sec during vigorous muscular exercise. *Gas exchange* between the alveolar air and the capillary blood by the process of physical diffusion must take place within this short interval of time. The process *takes place in steps:* through a number of different series-connected sections shown in figure 1.

Oxygen or carbon monoxide migrates from the alveolar space to the outer surface of the capillary epithelium in the first step. However, the resistance to gas diffusion in this step is negligibly small due to rapid movement of alveolar air. The gas then diffuses at the same rate through the pulmonary membrane which includes all the tissues interposed be-

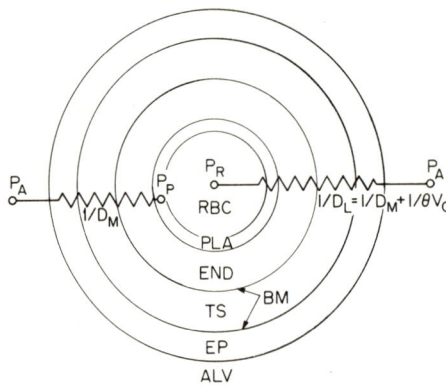

ALV = Alveolus
EP = Epithelium
BM = Basement Membranes
TS = Interstitial Tissue
END = Capillary Endothelium
PLA = Plasma
RBC = Red Blood Cell

Pulmonary Membrane

Fig. 1. Schematic diagram of cross section of a capillary in the human alveolar wall, between alveolar air and the red blood cell.

tween alveolar air and the plasma; that is, epithelium, basement membrane, interstitial tissue, basement membrane, and capillary endothelium. The diffusing capacity of the pulmonary membrane D_M is equivalent to the reciprocal of the pulmonary resistance $1/D_M$. The rate of diffusion is $N = D_M^*(P_A - P_P)$, where P_P is the partial pressure in the plasma outside the red cell and P_A is the partial pressure of the alveolar gas. D_M^* equals $(\alpha DA/x)_M$ where the subscript M refers to the material composing the membrane. Since the pulmonary membrane is not uniform in dimensions or composition, D_M^* must be a kind of average for the entire lung. There are no methods of estimating α, D, A and x independently.

After passing through the membrane, the gas diffuses from the plasma into the red cells in the capillary blood. The corresponding resistance is $1/\Theta V_c$ in which Θ is the rate of gas uptake in ml/(min · mm Hg · ml blood) and V_c is the pulmonary capillary blood volume in ml. Hence, *the diffusing capacity of the lung*, namely the total resistance between alveolar air and the interior of the red cell, D_L^*, can be expressed as

$$1/D_L^* = 1/D_M^* + 1/\Theta V_c. \tag{2}$$

One may now determine the rate of gas exchange in terms of the pressure difference $(P_A - P_R)$ between alveolar air and the red cell as

$$N = D_L^*(P_A - P_R). \tag{3}$$

The *change in the amount of dissolved gas* as the blood traverses the lung capillaries can be calculated through the following formulation.

One assumes that (1) the alveolar capillaries are uniform tubes with constant cross-sectional area A_c [cm²] and (2) the blood is well mixed across the capillary cross section at any location x measured from where the venous blood enters the capillaries. Taking a differential volume of blood $A_c \, dx$ as the control volume in figure 2, mass balance requires that

$$A_c \, dx \frac{d(\alpha_B P_c)}{dt} = (P_A - P_c) \frac{(\alpha D)_M \bar{P} \, dx}{l}$$

or, using the average values of α_B and D_M^* of the entire lungs, as

$$\alpha_B V_c \frac{dP_c}{dt} = D_M^*(P_A - P_c), \tag{4}$$

where P_c and P_A are the gas partial pressures [mm Hg] in the capillary blood at a location x [cm] and in the alveolar gas, respectively; t denotes the time [sec]; l represents the thickness of the capillary wall. The integration of equation 4 subject to the initial condition $P_c = P_{c0}$ at t (or x) = 0 yields

$$\frac{P_A - P_c(t)}{P_A - P_{c0}} = \exp\left(-\frac{D_M^* t}{\alpha_B V_c}\right). \tag{5}$$

Here, $x = ut$ in which u is the velocity of the blood stream in the capillaries. Equation 5 predicts the rise in the partial pressure of an inert gas in the blood as it moves through the capillary. Figure 3 shows the rate of oxygen uptake in the lung capillaries.

B. The Carriage of Oxygen and Carbon Dioxide in the Blood

The *principal functions of the circulating blood* are (1) the carriage of oxygen to the tissue cells and (2) the carriage of the waste products of cellular activity such as excess carbon dioxide to the lung capillaries for discharge in the expired gas.

Oxygen is mainly carried in combined form with hemoglobin, although it is soluble in plasma. In the lungs, oxygen molecules first diffuse across the alveolar-capillary membrane to be dissolved in the plasma of the capillaries. The oxygen that dissolves in the plasma then diffuses into red cells to combine with their hemoglobin. As the oxygenated hemoglobin reaches the tissue capillaries, the gas first dissolves in the plasma and is then transferred to the cells by simple diffusion. While 100 ml of plasma contains only about 0.3 ml of dissolved oxygen, the same volume of whole blood can carry approximately 20 ml of oxygen.

The oxygenation of hemoglobin in the lungs and the reduction of oxyhemoglobin in the tissues can be represented as

$$Hb + O_2 \rightleftharpoons HbO_2. \tag{6}$$

The forward and reversible reactions are equally rapid, taking less than 0.01 sec to complete, as compared to 0.3–0.75 sec for blood to traverse the lung capillaries.

The characteristics of the hemoglobin-oxygen combination vary with physiological conditions. This is shown in figure 4, which is called the *oxygen-hemoglobin dissociation curve*, at normal body pH of 7.4 and temperature of 38 °C. The curve shows the percentage saturation of hemoglobin with oxygen at each oxygen pressure level in the physiological range. The percentage saturation is based on the oxygen capacity of the hemoglobin, i.e., 20 ml of oxygen in 100 ml of blood as 100% saturation with oxygen. Oxygenated blood leaving the lungs has an oxygen pressure of about 100 mm Hg, at which approximately 97% of the hemoglobin is combined with oxygen. The oxygen pressure in the venous blood returning from the tissue capillaries falls to about 40 mm Hg. At

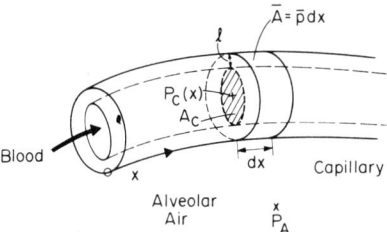

Fig. 2. Schematic of blood flow in pulmonary capillary vessel.

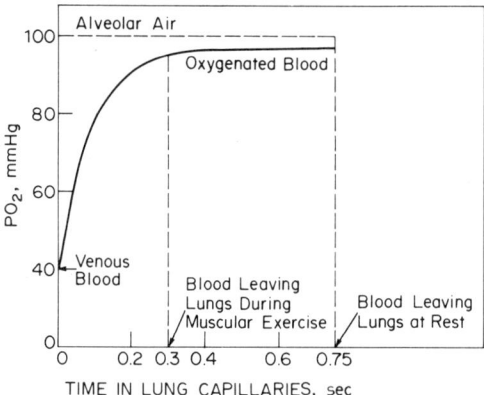

Fig. 3. Rate of oxygen uptake in lung capillaries.

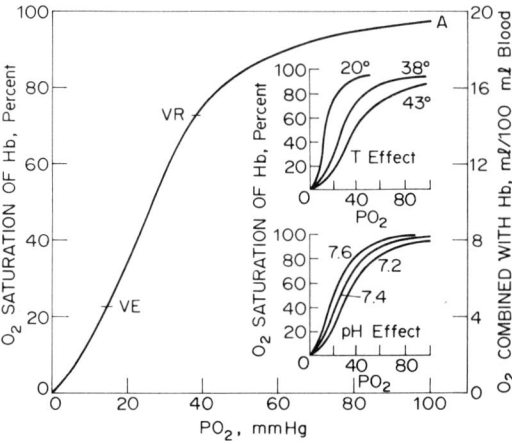

Fig. 4. The oxygen hemoglobin dissociation curve and effect of temperature and pH.

this pressure only about 70% of the hemoglobin is combined with oxygen. Thus, approximately 27% of the hemoglobin has lost its oxygen to the tissue cells. In severe muscular exercise, the PO_2 in the tissues may fall to as low as 20%, increasing the amount of oxygen transported to tissues as much as 3-fold. This increase in the arterio-venous oxygen difference is generally accompanied by a profound increase in the cardiac output as much as 6-fold. Then, the amount of oxygen that can be transferred to the tissue can be increased to as much as 21 times normal.

It is important to note that the dissociation curve may shift its position due to a change in temperature or pH, as shown in figure 4. As the pH value decreases to 7.2 due to the addition of carbon dioxide or lactic acid as in strenuous muscular activity, the curve shifts to the right. This shift causes oxyhemoglobin to release more of its oxygen. For example, when muscular exercise is performed, more carbon dioxide and other acids are formed, resulting in a decrease in the pH value and in turn releasing more oxygen from the hemoglobin. The shift of the curve is most profound in the steepest part. Consequently, the increase in oxygen demands is met and also the PO_2 in capillary blood is maintained at high value for efficient oxygen diffusion into the muscle cells. An increase in body temperature, resulting from muscular activity, for example, also shifts the curve to the right. In conclusion, an increase in temperature together with a decrease in pH would cause oxyhemoglobin to release more oxygen.

When oxygen is consumed by the cells for metabolism, carbon dioxide is formed. Carbon dioxide is readily soluble and dissolves as it is formed in the tissue cells:

$$CO_2 + H_2O \rightleftharpoons H_2CO_3 \rightleftharpoons H^+ + CO_3^-. \tag{7}$$

The reactions are reversible. The PCO_2 in the cells is about 55 mm Hg, while that of the blood entering the tissue capillaries is approximately 40 mm Hg. The pressure difference causes CO_2 to diffuse out of the cell into the interstitial fluid and then into the blood capillaries. The PCO_2 of the blood entering the lungs is about 45 mm Hg, while that in the air of the alveolus is only 40 mm Hg. This causes CO_2 to diffuse out of the blood, and the PCO_2 falls to 40 mm Hg in the arterial blood leaving the lungs.

Approximately 5% of CO_2 is carried in the blood in the dissolved state (in the plasma), while about 95% of CO_2 diffuses from the plasma into the red cell. In the plasma and interstitial fluid, the reaction

$$CO_2 + H_2O \rightleftharpoons H_2CO_3 \tag{8}$$

is slow and the dissociation of carbonic acid is slight. *In the red cells, however, CO_2 undergoes two reactions:*

(i) Approximately 65% hydrates through the presence of the enzyme catalyst, carbonic anhydrase. The reaction of equation 8 is fast and reversible. The carbonic acid immediately reacts with the acid-base buffers of the red blood cells to form hydrogen and bicarbonate ions. This reaction sequence can be described by equation 7. When hydrogen ions begin to accumulate in red cells, reduced hemoglobin acts as an hydrogen acceptor:

$$H^+ + Hb^- \rightleftharpoons HHb. \tag{9}$$

Thus, the accumulation of H^+ is prevented and the uptake of CO_2 continues. Clearly, as oxyhemoglobin dissociates to free oxygen for the tissue cells, the resulting reduced hemoglobin accepts hydrogen ions to further facilitate CO_2 uptake without any change in the corpuscular pH value. Simultaneously, bicarbonate ions migrate outwards to the plasma and the negatively charged ions in the plasma, mainly chloride ions, move into the red cells.

(ii) About 30% of the total CO_2 entering the blood combines directly with hemoglobin to form a compound called carbaminohemoglobin:

$$CO_2 + HbNH_2 \rightleftharpoons HbNHCOOH. \tag{10}$$

The reaction is reversible and very rapid, requiring no special enzyme.

When the blood enters the pulmonary capillaries, all the chemical combinations of CO_2 with blood are reversed, releasing the carbon dioxide into the alveoli.

III. Gas Emboli in Human Body

A. Sources of Gas Emboli

The sources of gas emboli may be classified into two main types depending on the motivations that bubbles are formed or introduced into the body.

1. Accidentally Introduced

A comprehensive survey of the literature pertinent to the human-body gas-embolism problems is presented by CHAN and YANG [1969b]. The sources of gas or air emboli include the following.

a) Decompression (the earliest known cases of air embolism which are associated with deep sea diving): during sudden decompression, the dissolved nitrogen (about 1 liter) forms bubbles throughout the body, particularly in the fatty tissues in abundance. This phenomenon has been known as *decompression sickness*, the bends, diver's paralysis, and others.

b) Nitrous oxide concentration. Nitrous oxide is 34 times more soluble than air in blood, giving rise to high nitrous oxide tension in the blood. Consequently, where there are nuclei of bubbles in the blood, nitrous oxide diffuses out into the bubbles, thereby increasing their size.

c) During general surgery. Atmospheric air is sometimes inadvertently introduced into the body for therapeutic and diagnostic procedures.

d) During accidents. Cases of air embolism arising from accidents involve rupture of veins, which offer an entrance to air.

e) Extracorporeal blood oxygenation. Air embolism (occlusion of a blood vessel by a gas embolus) resulting from extracorporeal blood oxygenation is one of the major concerns in open-heart surgery and heart transplant. Large areas of body surfaces normally covered with blood are brought in contact with air. Therefore, following the operation, some air bubbles are found to adhere themselves to the previously exposed surfaces. Minute and/or micro bubbles of air and oxygen in the extracorporcal circulation may be carried into the body and result in gas embolization.

f) Gas embolism following hypothermia. It arises in surgery during the warm-up period following deep hypothermia. Owing to the fall of gas solubility, the high concentration of oxygen or air in the cold blood gives rise to gas evolution during the warming-up period.

g) Fracture of bones and tissues. Bubble formation occurs in the regions of bone fracture and tissue damage due to intense mechanical agitation of the tissues.

2. Purposely Introduced

Sometimes, gas bubbles are purposely injected into subcutaneous tissues, liver and perirenal fat. Quantitative data obtained with the subcutaneous gas pocket, an *in vivo* tonometry[1] system, provide basic information concerning tissue-capillary gas exchange, specifically for the determination of (i) permeation of the dissolved gases in the tissues and

---

[1] Tonometry is a technique for determining gas permeability in living tissue-capillary systems.

their interaction with hemoglobin; (ii) tension of dissolved gases in the tissues in the constant composition state; and (iii) the influx and efflux of inert gases across the surface of a decomposition bubble. A brief summary on the literature pertinent to the problems of artificially-introduced gas pockets in tissues is presented by YANG and LIANG [1972].

B. Expansion and Dissolution of Gas Emboli in Blood

Gas emboli may grow or shrink due to mass diffusion, heat transfer and liquid inertia. Heat transfer mechanism is ignored since temperature difference in the body is usually small. *Theoretical models* have been developed to determine the rate of expansion or dissolution of gas emboli in blood for two cases depending upon the controlling mechanism: (i) mass-diffusion dominating and (ii) liquid-inertia controlling. Experimental studies have been performed to measure the dissolution of gas emboli in blood and plasma including the effects of blood flow velocity, plasma substitutes and anesthetics on the dissolution rate.

1. Theory

a) The Case Where Mass Diffusion Controls

As mentioned in section II.B., the dissolution of oxygen gas in blood is accompanied by simultaneous chemical reactions between the dissolved oxygen and the blood. The chemical reaction of second order takes place between the oxygen and the hemoglobin. The *combined processes of diffusion and chemical reaction* can be described by the following differential equation [FORSTER, 1964]:

$$\frac{\partial [O_2]}{\partial t} = D\nabla^2 [O_2] + k''[HbO_2] - k'[O_2][Hb] \tag{11}$$

where ∇^2 is the Laplacian operator; $[O_2]$, $[Hb]$ and $[HbO_2]$ are the concentrations of O_2, reduced hemoglobin, and oxygenated hemoglobin, respectively. When the chemical constituents are not in equilibrium, a second equation is required to define the relationships among the reactants [FORSTER, 1964; GIBSON, 1959]:

$$\frac{\partial [HbO_2]}{\partial t} = -k''[HbO_2] + k'[Hb][O_2]. \tag{12}$$

Numerical solutions of equations 11 and 12 have been obtained for several cases: (a) Diffusion of an oxygen gas into and out of an infinite sheet of hemoglobin solution, the one-dimensional case with and without a membrane (containing no hemoglobin) on the surface, by KLUG et al. [1956] and NICOLSON and ROUGHTON [1951]. (b) Diffusion of oxygen gas into spherical and discoidal volumes (three-dimensional) containing no hemoglobin, by FORSTER and VAN DE LINDT [1959].

If only the association of the dissolved gas and the reduced hemoglobin takes place, then equation (12) is not required while equation (11) may be modified as

$$\frac{D[O_2]}{Dt} = D\nabla^2[O_2] - K[O_2] \tag{13}$$

where K is k' [Hb] and D/Dt is the substantial derivative including both the local and convective effects on the diffusion.

Consider the *simultaneous diffusion and chemical reaction process* for the case where a gas contained in a spherical bubble diffuses into whole blood. The whole blood is treated as a homogeneous mixture of plasma and hemoglobin. If the reaction products such as oxygenated hemoglobin are constantly removed from the immediate vicinity of the gas bubble, then no dissociation process occurs and the gas bubble is surrounded by the reduced hemoglobin at a constant concentration. The situation fits two cases reasonably: (1) when the gas bubble is situated in a moving blood stream as observed in extracorporeal circulation, whence a relative motion occurs between the bubble and the blood, resulting from their bulk motion and the motion of the bubble surface due to its shrinkage, and (2) during the initial stage of bubble collapse in the quiescent blood.

Let R_0 be the initial radius of a spherical gas bubble situated in whole blood, in which the concentration of the dissolved gas is uniform and equal to C_∞. The liquid is at constant temperature T and pressure, and the dissolved gas concentration for a saturated liquid at this temperature and pressure is denoted by C_s. The origin of a spherical polar coordinate system is fixed at the center of the gas bubble. At any time t when the bubble radius is R, the dissolved gas concentration C at a point in the liquid at a distance r from the origin is governed by the diffusion equation (13). Since the convective effect on diffusion is small [YANG, 1971], it can be rewritten in spherical coordinates as

$$\frac{\partial C}{\partial t} = D\nabla^2 C - KC \tag{14}$$

where K>0 for whole blood, while K=0 in plasma where chemical reaction is absent.

The appropriate *initial and boundary conditions* are

$$C(r, 0) = C_\infty \tag{15}$$

$$C(\infty, t) = C_\infty, \quad C(R, t) = C_s. \tag{16}$$

Through the transformation

$$\zeta = r(v - C_s) \quad \text{and} \quad \xi = r - R, \tag{17}$$

the *solution of the problem* is found to be

$$C(r, t) = C_\infty + \frac{R(C_s - C_\infty)}{r} e^{-Kt} \left\{ 1 - \frac{2}{\pi} \int_0^\infty e^{-D\lambda^2 t'} \frac{\sin[(r-R)\lambda]}{\lambda} d\lambda \right\}$$

$$+ K \int_0^t e^{-Kt'} \frac{R(t')(C_s - C_\infty)}{r} \left\{ 1 - \frac{2}{\pi} \int_0^\infty e^{-D\lambda^2 t} \frac{\sin[(r-R)\lambda]}{\lambda} d\lambda \right\} dt'. \tag{18}$$

The concentration gradient at the bubble surface is

$$\left(\frac{\partial C}{\partial r}\right)_{r=R} = (C_\infty - C_s)\left(\frac{1}{R} + \frac{e^{-Kt}}{\sqrt{\pi Dt}} + \sqrt{\frac{K}{D}} \, \text{erf} \sqrt{Kt}\right). \tag{19}$$

Thus, the rate of mass flow into the liquid has the value

$$\frac{dm}{dt} = 4\pi R^2 D \left(\frac{\partial C}{\partial r}\right)_{r=R}. \tag{20}$$

However, the rate change of the mass of gas inside the bubble is

$$\frac{dm}{dt} = 4\pi R^2 \rho_g \frac{dR}{dt} \tag{21}$$

wherein ρ_g is the gas density. Hence, the combination of equations 19, 20 and 21 gives the differential equation for the bubble radius in dimensionless form as

$$\frac{dR^*}{dt^*} = -C_p^*(1 - C_\infty^*)\left(\frac{1}{R^*} + \frac{e^{-K^*t^*}}{\sqrt{\pi t^*}} + \sqrt{K^*} \, \text{erf} \sqrt{K^*t^*}\right)$$

$$\cong -C_p^*(1 - C_\infty^*)\left(\frac{1}{R^*} + \sqrt{K^*}\right) \cdots \text{near complete collapse} \tag{22}$$

in which

$$R^* = \frac{R}{R_0}, \quad t^* = \frac{Dt}{R_0^2}, \quad C_p^* = \frac{C_s}{\rho_g}, \quad C_\infty^* = \frac{C_\infty}{C_s}, \quad K^* = \frac{KR_0^2}{D}. \tag{23}$$

Equation 22 is integrated to yield the radius-time relationship for the bubble:

$$1 - (R^*)^2 = \tfrac{1}{2} C_p^*(1 - C_\infty^*)\left(t^* + \int_0^{t^*} \frac{R^* e^{K^*t^*}}{\sqrt{\pi t^*}} dt^* + \sqrt{K^*} \int_0^{t^*} R^* \, \text{erf} \sqrt{R^*t^*} \, dt^*\right). \tag{24}$$

The effect of surface tension (σ) on diffusion becomes important as bubble size decreases, especially near its complete collapse. The force balance on the bubble surface requires that

$$P + \frac{2\sigma}{R} = P_{g0}\left(\frac{\rho_g}{\rho_{g0}}\right)^n \tag{25}$$

where P and P_g are, respectively, the liquid and gas pressures exerted on the bubble surface, and the subscript 0 indicates the physical state corresponding to $t \leq 0$. The thermodynamic equation of state for an ideal gas gives

$$P = \rho_{g\infty} \frac{\bar{R}_g}{M} T. \tag{26}$$

Here, $\rho_{g\infty}$ is the gas density under the same conditions of pressure and temperature, with a gas-liquid surface of zero curvature. When equations 25 and 26 are incorporated, the bubble-dynamics equation 22 becomes

$$\frac{dR^*}{dt^*} = -\frac{C_\rho^*(\gamma^{1/n} - \rho^* C_\infty^*)}{\gamma^{1/n}(1 - \sigma^*/3R^*n\gamma)} \left(\frac{1}{R^*} + \frac{e^{-K^*t^*}}{\sqrt{\pi t^*}} + \sqrt{K^*} \operatorname{erf} \sqrt{K^*t^*}\right) \tag{27}$$

in which

$$\gamma = \rho^* + \frac{\sigma^*}{R^*}, \quad \rho^* = \frac{\rho_{g\infty}}{\rho_{g0}}, \quad \sigma^* = \frac{2\sigma}{R_0 P_{g0}}.$$

It has been revealed through an examination of equations 22 and 27 that (1) when the effect of surface tension is neglected, the radius-time relation for dissolving bubbles and, consequently, the diffusion of the dissolved gas in the liquid are functions of *four dimensionless parameters* C_ρ^*, C_∞^*, K^* and t^*; (2) when the effect of surface tension is taken into consideration, both the diffusion and the radius-time relation become functions of *seven dimensionless parameters* C_ρ^*, C_∞^*, K^*, t^*, σ^* ρ^* and n. The parameters ρ^* and n depend on the thermodynamic process of the gas inside the bubble during the bubble collapse. For an isothermal process, both ρ^* and n are unity and, consequently, only the remaining five parameters influence the diffusion and the bubble dynamics.

Numerical results are obtained for $C_\infty^* = 0$ using a digital computer. Figure 5 shows the concentration-time relation for both whole blood ($K^* = 4,880$, corresponding to $K = 9.76 \text{ sec}^{-1}$, $D = 2 \times 10^{-5} \text{ cm}^2/\text{sec}$ and $R_0 = 0.1$ cm, for example) and plasma ($K^* = 0$). r^* is defined as r/R_0. It is seen that concentration boundary layers in the solution around the bubble through which the diffusion takes place become larger than the bubble itself for small times. This observation confirms that the omission of the convective term in equation 14 is justified. The radius-time relations for

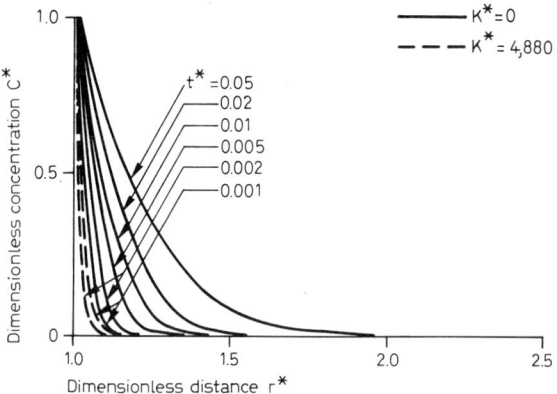

Fig. 5. Concentration-time relation for dissolved gas in whole blood ($K^* = 4,880$) and plasma ($K^* = 0$) for $C_\infty^* = 0$.

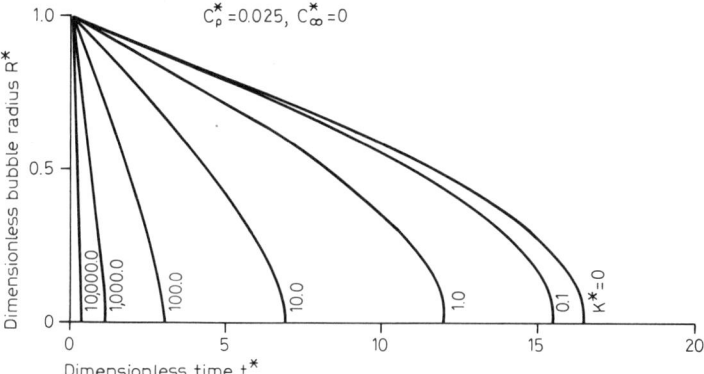

Fig. 6. Radius-time relation for dissolving bubbles in whole blood for $C_\rho^* = 0.025$ and $C_\infty^* = 0$.

dissolving bubbles in whole blood with $C_\rho^* = 0.025$ (corresponding to $C_s = 3.26 \times 10^{-5}$ g/cm³ and $\rho_{g0} = 1.307 \times 10^{-3}$ g/cm³) and in plasma are depicted in figures 6 and 7, respectively.

Figure 6 illustrates that as the magnitude of the reaction velocity constant K^* is increased, the time of complete dissolution of a gas bubble in the whole blood is shortened. This observation can be easily explained physically, since the term $-KC$ on the right-hand side of equation 14 signifies the existence of a uniformly distributed mass sink of strength KC in the liquid region; the larger the value of K, the stronger is the strength of the mass sink, and consequently the diffusion is enhanced. This results in a faster dissolution of a gas bubble in the liquid.

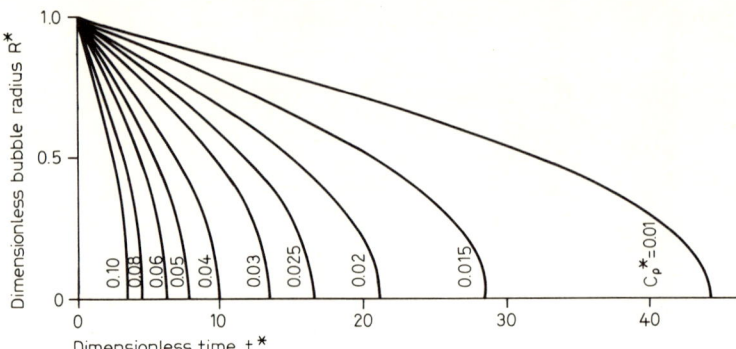

Fig. 7. Radius-time relation for dissolving bubbles in plasma ($K^* = 0$) for $C_\infty^* = 0$.

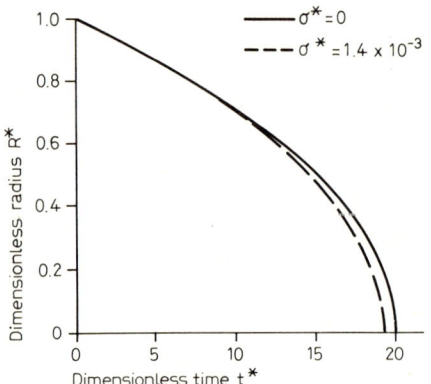

Fig. 8. Effect of surface tension on radius-time relation for dissolving bubbles in whole blood with low K^* and plasma for $C_\rho^* = 0.025$ and $C_\infty^* = 0$.

Figure 7 indicates that the time required for a gas bubble to dissolve completely in the plasma is shortened as the value of C_ρ^* increases, i.e., as C_s is increased and/or ρ_g is decreased. This means that at the same temperature at saturation, the bubble of a gas with lower density may be completely dissolved in the plasma in less time than that of a gas with higher density.

The effect of surface tension on the bubble collapse (in whole blood) is illustrated in figure 8. $\sigma^* = 1.4 \times 10^{-3}$ corresponds to $\sigma = 70$ dyn/cm, $R_0 = 0.1$ cm and $P_{g0} = 1.00140 \times 10^6$ dyn/cm^2. It is seen in the figure that surface tension helps to shorten the time of complete solution, i.e., the larger the surface tension force, the faster will be the bubble collapse.

Through approximations valid for a saturated solution, $C_\infty^* = 1$, the time of complete dissolution is obtained from the solution of equation 27 with $n = 1$, $\rho^* = 1$, $K^* \to 0$ and large values of t^* as

$$t_f^* = \frac{1+\sigma^*}{3\sigma^* C_p^*} \tag{28}$$

which decreases with an increase in surface tension σ^*. t_f is called the bubble lifetime.

For determining the hydrodynamic mechanism of blood trauma later in the chapter, it is necessary to derive the expressions which predict the maximum liquid pressure and its location: mass continuity requires that the radial velocity at any distance r from the bubble center is

$$u(r, t) = \dot{R}(R/r)^2. \tag{29}$$

Integrating the Navier–Stokes equation of motion, with respect to r, from r to infinity, it yields

$$P - P_\infty = \frac{\rho}{g_c}\left(\frac{Q}{R}\right)^2\left(\frac{S}{Rr} - \frac{R^2}{2r^4}\right) \tag{30}$$

in which g_c is the conversion factor, equivalent to $1.0 \, g \cdot cm/(dyn \cdot sec^2)$, and

$$S = \frac{1+2K^{1/2}R}{1+K^{1/2}R}. \tag{31}$$

The maximum liquid pressure (i.e., shockwave front), P_{max}, is then found from equation 30 as

$$P_{max} = P_\infty + \frac{3\rho}{2^{1/3}g_c}\left[\frac{Q(1+K^{1/2}R)}{2R}\right]^2 S^{4/3} \tag{32}$$

which occurs at

$$\gamma = 2^{1/3}R/S^{1/3}. \tag{33}$$

If the effect of surface tension on diffusion is negligible, *the bubble lifetime* can be obtained from equation 24 for small values of K^* and large values of t^* (retaining only the t^* term in the parentheses on the right-hand side) as

$$t_f = \frac{2\rho_g}{D(C_s - C_\infty)}. \tag{34}$$

b) The Case Where Liquid Inertia Controls

In the previous case, the concentration at the bubble wall quickly approaches the saturation value corresponding to the external pressure, and hence the pressure difference in the liquid at the bubble wall and at a point far away from the bubble becomes essentially zero. The liquid inertia may thus be neglected and the dissolution is controlled by the rate of mass diffusion from the bubble to the liquid. On the other hand, conditions may be such that mass transfer effects are minor and liquid inertia may be expected to play a dominant role. In general, however, there is a coupling between the momentum equation and the equation for the concentration field in the liquid, and the problem then becomes quite complex.

Behavior of gas emboli controlled by liquid inertia mechanism in blood has been studied by CHAN and YANG [1969a]. It is disclosed that *a Newtonian fluid model may be used for blood in the analytical study of dynamic behavior of gas emboli situated in the blood.* The so-called bubble dynamic equation reads

$$\rho(R\ddot{R} + \tfrac{3}{2}\dot{R}^2) = P_g(R) - P(\infty) - \frac{2\sigma}{R} - 4\mu\frac{\dot{R}}{R}. \tag{35}$$

Here, R, $\dot{R}$, and $\ddot{R}$ represent the instantaneous bubble radius and its first- and second-time derivatives, respectively. The right-hand side of equation 35 represents the liquid inertia effect which may be induced by the pressure variation around the bubble $P_g(R) - P(\infty)$, surface tension $2\sigma/R$ and liquid viscosity $4\mu\dot{R}/R$.

The initial conditions describe the bubble initial size $R(0) = R_0$ and at static state $\dot{R}(0) = 0$. For the gas in the embolus, undergoing a reversible polytropic process, its pressure may be expressed in the form of equation 25, noting that

$$P_g(R) = P(R) + \frac{2\sigma}{R}.$$

The variation in the system pressure $P(\infty)$ is the forcing function. Two typical variations in $P(\infty)$ are considered:

$$P(\infty) = \begin{cases} P_b(1 + F\sin\omega t) \\ P_b(1 + F\sin\omega t + f) \end{cases} \tag{36}$$

corresponding to a sinusoidal change and a step change, respectively. Here, P_b is the mean blood pressure, $P_b F$ and ω are, respectively, the amplitude and frequency of blood pressure pulse, and $P_b f$ is the amount of a step change in the external pressure.

For a subject under normal atmospheric conditions, the blood pulse frequency is about 72 cycles/min, or natural frequency of an embolus of radius 0.01 cm in human blood is about 2×10^6 cycles/min. Therefore, *relative to the natural frequency of the embolus, the frequency of the blood pressure fluctuation is very low.* Under this condition, the embolus will oscillate at its natural frequency. This implies that as the blood pressure undergoes one cyclic change, the embolus oscillates as many times as its own natural frequency. Therefore, the effect of the blood pressure fluctuation on the embolus is like that of blood pressure slowly applied to the embolus. Hence, the instantaneous mean bubble radius can be determined using the equation

$$R = 2\sigma/[P_g - P(\infty)] \tag{37}$$

obtained by the equilibrium condition of static pressure, where $P(\infty)$ is the instantaneous blood pressure.

2. Experiments [YANG et al., 1971]

The theory presented in section III.B.1.a) is limited in applications to the cases where chemical reaction between the dissolved gas and the liquid is simple and straightforward. When the chemical reaction is complex, such as in the case of the dissolved carbon dioxide in plasma or blood, the theory becomes invalid and experiments must be performed.

The experimental study on the dissolution of a gas bubble in degassed blood and plasma is conducted in two steps. *The first step* is to degas the liquid. During this stage, the gas dissolved in plasma is removed and oxyhemoglobin in whole blood dissociates into reduced hemoglobin and oxygen which is removed as soon as it is formed. *The second step* involves the injection of a gas bubble into the degassed blood or plasma, followed by the measurements of bubble size at certain time intervals until the bubble is completely dissolved in the liquid.

The equipment used in the degassing process is shown in figure 9.

The test fluid is placed in a 250-ml separatory funnel with a thermometer inserted into the fluid. A Tygon tubing connects the funnel to a 250-ml distilling flask, which traps fluids being carried out of the funnel during the degassing process. A 250-ml Erlenmeyer flask retains any fluid being transported up the tubing from the distilling flask. The Erlenmeyer flask is connected to a T joint. One branch of the T leads to a mercury manometer, while the other branch is connected to a condensing trap which removes much of the water vapor from the air being drawn from the system. A vacuum pump is used to create a vacuum of approximately 28.5 in. of mercury in the system during the degassing process.

When the degassing is complete, the separatory funnel is placed on a wooden support. Using a syringe with a size No. 20 needle, a gas bubble is introduced into the degassed liquid in the separatory funnel. The illumination is provided by two 15-W white fluorescent lamps. A reticle with long metric scale is placed on the side of the funnel. A dual eyepiece

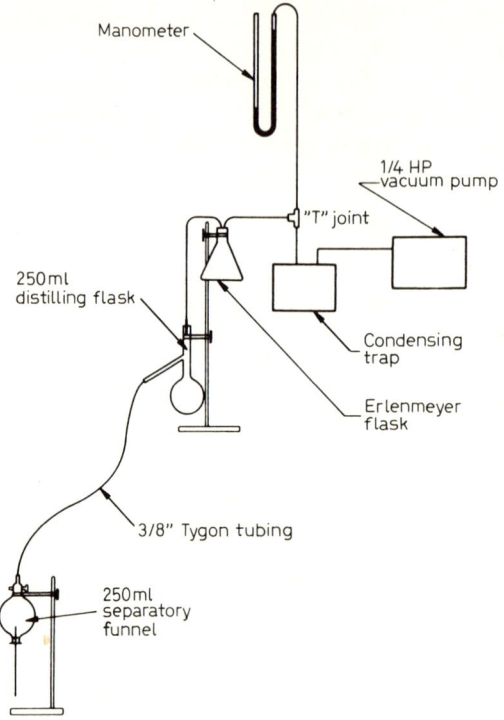

Fig. 9. Apparatus for degassing process.

microscope is placed above the funnel and reticle for reading the bubble size.

Possible deterioration of the biological fluid resulting from the degassing process or its continuous exposure to the room temperature and pressure is checked by measuring a change in liquid viscosity before and after each bubble test: the test liquid is allowed to flow through a capillary tube into a flask. The time required for the liquid to fill the flask is recorded as used as a proportional change in liquid viscosity.

Experiments were conducted for the dissolutions of oxygen, nitrogen and carbon dioxide bubbles in the degassed blood and plasma. Some typical results indicating the radius-time relation are illustrated in figures 10 and 11. When a carbon dioxide bubble is injected in the degassed plasma, a concentration boundary layer of the dissolved CO_2 is formed around the bubble. As mentioned in section II.B., the carbon dioxide quickly combines with water to form carbonic acid. This is a loose, reversible combination due to the reversible formation of bicarbonate ions as described by equation 7. The rapid dissolution of carbon dioxide

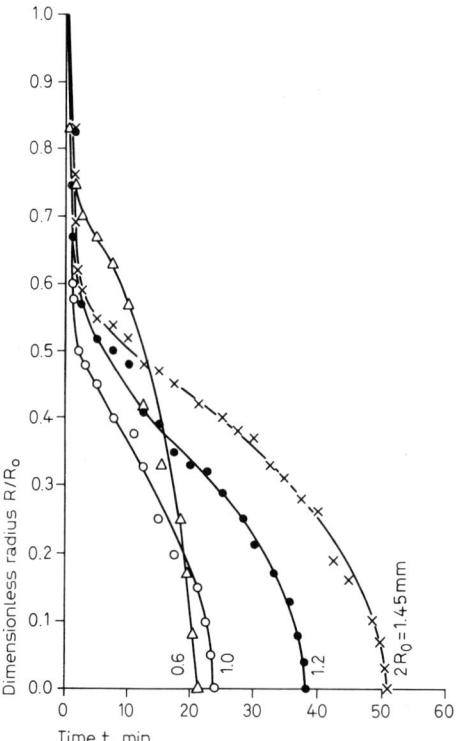

Fig. 10. Radius-time relation of carbon dioxide bubbles in degassed plasma.

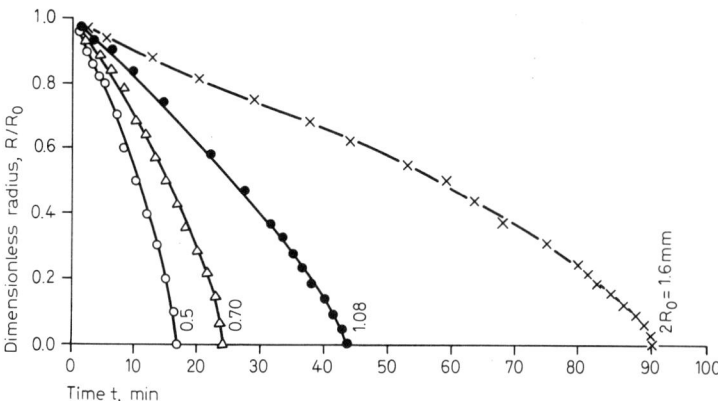

Fig. 11. Radius-time relation of oxygen bubbles in degassed plasma.

bubbles shown in figure 10 indicates the occurrence of chemical reaction immediately following injection. The subsequent shrinkage of the bubbles until complete dissolution is due to diffusion mechanism.

A similar radius-time relation exists in the dissolution of oxygen bubbles in degassed blood, except the chemical reaction of equation 6 to form oxyhemoglobin is much more rapid: an oxygen bubble shrinks so rapid immediately following injection into the blood that a difficulty is encountered in reading bubble size by naked eye due to the color of the blood. A movie camera is then employed to record the initial phase of bubble-size history [TANASAWA et al., 1971].

The dissolutions of oxygen (fig. 11) and nitrogen gases in the plasma are due to diffusion alone. One observation is that nitrogen gas bubbles have the longest lifetime in the plasma and consequently also in the blood, when compared to those of oxygen and carbon dioxide. One should bear in mind that the speed of dissolution depends on the magnitude of diffusion coefficient and solubility. Another important observation is that the rate of shrinkage of bubbles is accelerated near the completion of dissolution as evidenced by the steeper slope of the radius-time curves – resulting in the generation of shock waves and liquid jets to damage red blood cells as will be discussed later in section IV.D.4.

The experimental method, in combination with the theory of section III.B.1a), is also applied to determine the mass diffusivity (diffusion coefficient) of gases in plasma and the reaction velocity constant with hemoglobin in intact red cell suspensions in human and dog bloods [TANASAWA et al., 1971].

C. Effects of Foreign Agents on the Behavior of Gas Emboli

The presence of a foreign agent in the blood may cause a change in certain physical properties such as the coefficient of diffusion, the surface tension and viscosity of the blood, and others. One is then tempted to use a harmless foreign agent which possesses such characteristics that its presence may change the physical properties in favor of accelerating the dissolution of the gas embolus. Two important foreign agents are considered: plasma substitutes and anesthetics. They are used extensively not only in extracorporeal blood oxygenation processes but also in many other surgical procedures.

1. Plasma Substitutes

Dextrans are effective plasma substitutes used in the prevention or treatment of hypovolemic shock and as isotonic solution to prime pump

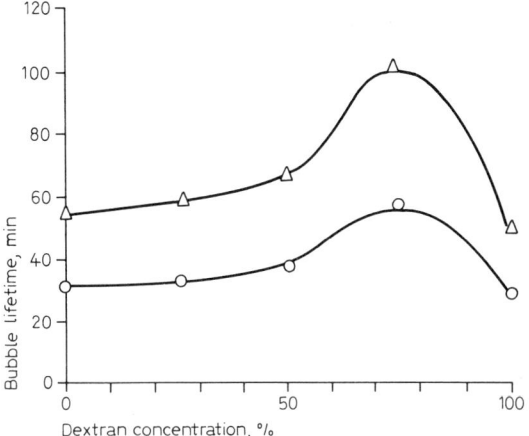

Fig. 12. Effect of dextran on lifetime of oxygen bubbles in whole blood for $R_0 = 0.4$ min ($\triangle$) and $R_0 = 0.3$ mm ($\bigcirc$).

or to improve flow in surgery requiring cardiopulmonary bypass, as well as to reduce the viscosity of blood and thereby enhance tissue perfusion.

Experiments were performed to determine *the effects of dextran on the dissolution of oxygen bubbles in degassed blood* using the equipment and procedure described in section III.B.2. [YANG and MA, 1973]. The dissolutions of oxygen bubbles are tested in several mixtures of different compositions of dextran 70 and the whole blood. Figure 12 shows the bubble lifetime versus the concentration of dextran in the whole blood. It is seen that *the lifetime of oxygen bubbles is increased in the presence of dextran*. This increase in the bubble lifetime is most significant, up to almost twice that in the whole blood, at the dextran concentration of 70–80%. The observation signifies that gas embolism resulting from extracorporeal oxygenation of the blood would be less acute when a low concentration of dextran is presented in the blood.

2. Anesthetics

It has been shown in laboratory as well as clinical applications that air-filled cavities within the body may increase in volume and/or pressure during nitrous oxide anesthesia [MUNSON and MERRICK, 1966]. This phenomenon has been attributed to the 34-fold difference in blood solubility between nitrous oxide and nitrogen. Of course, it is related to the difference in the diffusion coefficients of nitrous oxide and nitrogen dissolved in blood, as remarked in section III.B.2. This results in more rapid transfer of nitrous oxide molecules from blood (which contains a high partial pressure of nitrous oxide and essentially no nitrogen) into an air-filled cavity (which contains a high partial pressure of nitrogen) than removal of nitrogen molecules from the air space. The same principles of gaseous exchange apply to any gas of sufficient solubility in blood (relative to nitrogen) present in blood at high partial pressure, for example, halothane.

The increase in volume attainable by an air space is related to the concentration (partial pressure) of the foreign agent in the blood to which it is exposed. In case of nitrous oxide, MUNSON, and MERRICK [1966] have observed a maximum increase of 4-fold at an alveolar (arterial) concentration of 75%. They have also calculated a maximum increase in gas volume of 3.4-fold if the venous blood were in equilibrium with a nitrous oxide concentration of 70%. Since a larger bubble takes a longer time to dissolve, as illustrated in equation 34, the bubble lifetimes are maximum at the nitrous-oxide concentrations of 75 and 70% in the arterial and venous blood, respectively. The observation supports the experimental result that the *oxygen lifetime takes a maximum value at a dextran concentration of* 70–80% *in blood.*

IV. Gas Embolism Due to Extracorporeal Oxygenation

A. Artificial Heart-Lung Machines

An artificial heart-lung machine for extracorporeal blood oxygenation consists of a pump (or two pumps in the perfusion circuit with a recirculating line as required for a membrane-type or screen oxygenator), a gas-exchange device called oxygenator or artificial lung, and subsidiary equipment. The pump is to be used to replace the function of a heart. Single- or double-roller pumps are generally satisfactory, while a variety of ventricle pumps are also available.

Gas exchange devices may be divided into two general types: nonmembrane and membrane types. In the nonmembrane-type devices, blood and gas are in direct contact and thus the gas exchange is most efficient. Three forms of the nonmembrane-type devices are commonly used: bubble, disc and screen (or film) oxygenators. The membrane-type devices have a plastic film interposed between blood and gas. The interposition of a membrane between the blood film and the gas reduces blood damage, but the efficiency of the gas exchange also suffers. Although several general varieties of membrane-type devices are available commercially, development work on these devices is still in progress.

The subsidiary equipment includes cannulas and tubing, heat exchangers, filters and bubble traps, reservoirs (including venous, priming, and coronary return reservoirs) and control devices.

The basic circuit is a series arrangement of venous (venous return control, reservoir), oxygenator and arterial (pump, heat exchanger, bubble trap and filter) elements. It serves for disc, bubble and some membrane-type devices. The screen oxygenator and most membrane-type devices require, in addition, a recirculating line.

B. Blood Trauma

Although thousands of cardiac patients are currently operated on each year, *the safe duration of total body perfusion by extracorporeal gas exchange devices* is limited by a host of pathological phenomena resulting from blood trauma, including red cell hemolysis, plasma protein denaturation, disturbance of coagulation factors and the risk of emboli. There is evidence that the presence of a direct blood-gas interface results in extensive denaturation and significant hemolysis.

The knowledge of hemolysis and denaturation is summarized by GALLETTI and BRECHER [1962], PIERCE [1969], and BERNSTEIN et al. [1967].

1. Possible Factors Responsible for Hemolysis

Most of the RBC destruction in open-heart procedures is caused by the pumps and constrictions in the circuit. Significant hemolysis also occurs in the nonmembrane-type oxygenators, especially of the bubble type. The extent of destruction depends on the material used for tubing and the pump-oxygenator parts. Generally, smoother walls produce less hemolysis. For any given tube size, there is a critical velocity beyond which hemolysis is produced. Excessive suction with foaming in the cardiotomy suction line is the most important single factor producing hemolysis in extracorporeal circulation procedures.

Intermittent positive pressure is more likely to produce hemolysis, while sustained high pressures applied to the blood are not determinantal. The shearing force of bubble formation may be responsible for hemolysis in bubble oxygenators with large gas flow through minute orifices. Shearing action on the squeezed red cells produces hemolysis in occlusive pumps, whereas turbulence causes blood damage in unocclusive pumps. Further, a recent study [YANG, 1974] has concluded that the impact pressures and shear stresses produced by shock waves and liquid jets formed by involution of collapsing gas and cavitation bubbles are responsible for hemolysis and denaturation.

Overheating of blood can cause severe hemolysis. The liability of hemolysis increases by storage of heparinized blood before use and decreases with dilution of the blood with physiologic solutions.

2. Denaturation

Denaturation, thought to be caused by the severe and unnatural surface tension forces encountered in the circuit, occurs in all types of extracorporeal devices. The reactive bounds of protein molecules are torn apart as these molecules are forced from a more globular to a linear configuration by the action of surface tension. Fat attached to proteins is freed, forming increasingly larger neutral fat aggregates.

C. Introduction of Gas Emboli into the Blood during Open-Heart Surgery

Gas emboli can be introduced into the blood either at the operative field or in the pump oxygenator. In open-heart surgery and heart transplants, large areas of body surfaces, normally covered with blood, are brought in contact with air. Therefore, following the operation, some air bubbles are bound to adhere themselves to the previously exposed surface. Besides, in returning the blood into the heart, air is invariably introduced together with it. In the *pump oxygenator,* there are three different ways gas emboli may be introduced into the blood: (1) when pockets of air remain behind during the filling of the blood circuit, they can be set free after the start of the perfusion; (2) the oxygen used in the bubble oxygenator produces many bubbles, although a defoaming section would remove these bubbles; (3) blood, mixed with air, is sucked from the open heart and reintroduced into the pump oxygenator.

By means of ultrasonic detectors, LUBBERS *et al.* [1974] have investigated the contributions of these different sources on the number of emboli in the blood leaving a pump oxygenator (roller type pump and Bentley oxygenator) through the arterial line to the patient. The duration of perfusions ranged from 27 to 145 min with blood flows of 1.0–4.0 liters/min (for a medium size oxygenator) and of 3.0–5.2 liters/min (for a large unit). It is disclosed that the amount of emboli leaving the machine depends mainly upon the PO_2 and the rate of blood flow. Emboli are generally smaller than 0.1 mm in diameter at the entrance to the arterial line and dissolve rapidly at low PO_2 values. Undoubtedly, emboli in the oxygenator and arterial elements must be larger than 0.01 mm diameter and more numerous. Air trapped in filters gives rise to much larger emboli. The amount of suction also affects the number of gas emboli only to a small degree. It is more effective in controlling the number of emboli maintaining PO_2 low, about 100 mm Hg, rather than using filters with small size pore, say 0.03 mm diameter.

D. Hydrodynamic Basis of Hemolysis

1. Sphering of Erythrocytes

Human red blood corpuscles are biconcave discs in appearance in the relaxed state (fig. 13). At equilibrium, the resultant of the surface tension and the elastic stress acting on the cell membrane must balance with the difference in pressures, including both the osmotic and hydrostatic components, across the cell membrane, ΔP: If the red blood cell is biconcave, ΔP is about 230 dyn/cm². However, ΔP can be altered by either a change in the surrounding fluid or due to a concentration gradient across the semipermeable cell membrane, thus producing an osmotic pressure difference.

The critical buckling pressures P_{crit} of biconcave RBC membrane are determined as 120 and -20 dyn/cm² for radii of curvature equal to 10 μm (at g in fig. 13) and 1.2 μm (at b), respectively [FUNG, 1966]. When ΔP exceeds P_{crit}, the biconcave disk configuration becomes unstable and the shape of the red blood cell deforms drastically. In lysins, it transforms to a reversible disk-sphere, with the cell volume remaining unchanged. If the lytic process continues, the cell becomes a prolytic sphere, which finally hemolyzes leaving its remains called 'ghost'.

2. Critical Membrane Yield Stress

Shear stresses on the cell membrane may be produced by the shearing motion in the blood stream due to either sudden dimensional (flow cross-sectional area) changes in tubing and cannulas or jets. In jet tests [BERNSTEIN et al., 1967], measurable hemolysis (cell rupture) commences at the jet velocity U of 2,000 cm/sec. The equivalent shear stress in the circular jet is [SCHLICHTING, 1960]

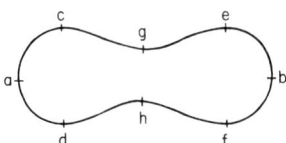

Fig. 13. Human red blood corpuscle in relaxed state.

$$\tau_{jet} = 0.015 \rho U^2 \tag{38}$$

or 6×10^4 dyn/cm². This means *the critical membrane yield stress for cell rupture is* 6×10^4 *dyn*. According to TAYLOR [1934], the shear stress at droplet breakup is related to the effective surface tension, σ (that is, the membrane yield strength), by

$$4\tau \frac{19\lambda^* + 16}{16\lambda^* + 16} = \frac{2\sigma}{\tau} \tag{39}$$

wherein λ^* is the ratio of suspended to suspending phase viscosity. The values of λ^* are 5 and 10 for cells in plasma and in saline, respectively. With $\lambda^* = 5$, one obtains the critical σ of 36 dyn/cm, or the critical surface-tension force $2\sigma/a$ of 1.8×10^5 dyn/cm², where a is the radius of a sphered RBC, 4×10^{-4} cm. The centrifugal test yields the critical cell membrane tension of 5–10 dyn·cm.

3. Wall and RBC Collisions

It has been known that wall-RBC interactions are responsible for most of the RBC destruction in open-heart surgery and that the material used for conduits affects the extent of cell destruction.

4. Shock Waves and Liquid Jets Produced by Collapsing Emboli [YANG, 1974]

Symmetrical bubble collapse produces shock waves. For a spherical gas bubble dissolving in blood due to mass diffusion, equations 22 and 32 give the radial velocity of contraction and the maximum liquid pressure (shockwave front), respectively. As will be shown later in the section, the compressive pressure pulse or shock front radiated from the collapsing motion of a bubble at some distance, equation 33, from the surface of a solid body is the source of blood damage. That is to say, if a red blood cell happens to be at or near the location of a shock wave, the liquid impact will damage the cell membrane.

Asymmetrical bubble collapse produces liquid jets. When gas bubbles are situated near solid walls (of tubing and the pump-oxygenator parts) or erythrocytes, then, under the influence of asymmetrical disturbances such as asymmetrical distributions of the surrounding liquid velocity and pressure, they may suffer shape instability, resulting in asymmetric collapse. Near its final stage, the involution of the collapsing bubble produces a liquid jet of high velocity. Using equation 22 to approximate the contraction velocity of the bubble, the instantaneous pressure produced by this liquid jet is

$$P_{jet} = \left(\frac{dR}{dt}\right)(\rho\beta)^{1/2} \tag{40}$$

wherein β is the coefficient of volume compressibility of blood. Near the conical boundary of this liquid jet, the shear stress given by equation 38 with $U = dR/dt$ is produced. A large destructive force produced by the impact of this liquid jet as well as a high shear stress due to the shearing motion of the liquid are the causes of RBC damage.

Numerical results are obtained for the *oxygen-blood system*: $K = 9.76 \text{ sec}^{-1}$, $D = 2 \times 10^{-5} \text{ cm}^2/\text{sec}$, $C_s = 3.26 \times 10^5 \text{ g/cm}^3$, $\rho_g = 1.307 \times 10^{-3} \text{ g/cm}^3$, $P_\infty = 1.013 \times 10^6 \text{ dyn/cm}^2$, $P_g = 1.014 \times 10^6 \text{ dyn/cm}^2$, $\rho = 1 \text{ g/cm}^3$, $C_\infty = 0$, and $\beta = 2.18 \times 10^{10} \text{ dyn/cm}^2$. For an oxygen bubble collapsing to $R = 10^{-3}$ cm (say, from $R_0 = 0.1$ cm), one finds $(P_{max} - P_\infty)$, P_{jet} and τ_{jet} of 6.5×10^{13}, 1.23×10^{11} and $1.25 \times 10^5 \text{ dyn/cm}^2$, respectively. If $R = 10^{-2}$ cm (say from $R_0 = 0.2$ cm) is used for calculations, $(P_{max} - P_\infty)$, P_{jet} and τ_{jet} are 2.02×10^{13}, 5.67×10^{10} and $5.77 \times 10^4 \text{ dyn/cm}^2$, respectively. The magnitudes of these pressures and shear stresses are sufficient to damage the red blood cells.

One may thus conclude that in the nonmembrane-type oxygenators, the impact pressures and shear stresses produced by shock waves and liquid jets formed by involution of collapsing oxygen bubbles are a major cause in blood trauma.

E. Gas Embolism Syndromes, Prevention and Treatment

Gas (air, in particular) embolism is a dreaded complication in surgical, therapeutic, and diagnostic procedures. A comprehensive survey of the literature related to gas embolism is available [CHAN and YANG, 1969b]: In one report, fatal venous air embolism accounted for one in 108 deaths in the operating room, while another report gave a rate of 2.6% for air embolism during occipital craniotomy. Bends associated with deep sea diving has been reputedly caused by air embolism on decompression from high pressures. The possibility of air embolism in open heart surgery, heart transplant and space travel is even greater.

1. Types of Air Embolism

Fundamentally, there are two main types of air embolism, namely pulmonary or venous and arterial. *Arterial air embolism* results from the entrance of air into the left side of the heart from the pulmonary veins or through septal defects. The air bubbles escape into the systemic and coronary arteries and the clinical manifestations depend on the site of arterial occlusion. *In the large arteries, the presence of air emboli is harmless.* The bubbles are carried along with the blood into the smaller vessels. When the bubbles are small, they are very 'hard'; and when they reach the capillary tubes of the arterial system, they are lodged in them. For this reason, even small amounts of air may cause considerable lasting damage. The occlusion of the small blood vessel by the bubble prevents blood from flowing, and this persists until the bubble is dissolved. Long-lasting bubbles, therefore, may cause damage similar to thrombosis and other clinical manifestations. Various degrees of cardiac damage may be caused by small amounts of air introduced directly into the coronary arteries (*in vivo* tests conducted on mongrel dogs). Similarly, cerebral air embolism due to blockage of small vessels in the brain may produce permanent gross neurologic defects in the survivors.

In the case of *pulmonary or venous embolism*, air enters a systemic vein and is carried to the right side of the heart. The churning action of the heart on the blood converts the air-blood mixture into a foam. If the foam is not excessive, it passes into the lungs where the air is filtered out. If the amount of air is more than the heart can eliminate in this manner, it accumulates in the heart resulting in failure of the right heart. However, before the onset of cardiac failure, there are various manifestations such as rise of pulmonary artery pressure, fall of systemic blood pressure and the characteristic drumming and millwheel sound. The hemodynamic alterations produced by air within the central venous pool, cardiac chambers and coronary vessels have been studied by many investigators [for example, MARCHAND *et al.*, 1964]. GOTTLIEB *et al.* [1965] tabulated the dosage of air that would prove fatal and showed the various possible situations where venous air embolism may occur and suggested their remedy.

Between these two types of air embolism, it seems venous embolism accounted for most of the deaths arising from air embolism. This may be because in arterial embolism, unless there is a septal defect, there is very little opportunity for air to pass through the filtering action of the lungs to go into the arterial system; in the venous system, there is no such protective mechanism to stop the air from entering the right heart. It may also be because arterial embolism is more difficult to diagnose; therefore, many such cases have not been detected.

Other types of air embolism, such as renal and hepatic, have been detected in experiments, but no clinical reports on their harmful effect are available.

The extent of damages caused by gas embolism depends also on the type of gas within the bubble and in the body fluid, as quantitatively discussed in section III.B. Carbon dioxide and oxygen, which have a higher solubility in blood, are better tolerated. On the other hand, air which consists mainly of nitrogen is quite injurious.

2. Treatment and Prophylactic Measures

The remedy for air embolism depends on the nature and circumstance of embolism.

In the case of air embolism in the circulatory system, preventive measures are widely recommended. These include changes of design in equipment which have so far proved to be giving rise to air embolism. GOTTLIEB *et al.* [1965] gave a detailed description of the equipment, their defects, and the recommended changes of design. The use of carbon dioxide gas to flood the operation field has been widely recommended for open heart surgery [SELMAN *et al.*, 1967; BURBANK *et al.*, 1965; EGUCHI *et al.*, 1963]. This method seems to be in popular use nowadays. Where air embolism is suspected, avoidance of nitrous oxide as an anesthetic is strongly advocated so as to prevent diffusion of nitrous oxide into the bubble nucleus [MUNSON and MERRICK, 1966].

However, in practice, air invariably enters the circulatory system in one way or the other. *In the case of pulmonary embolism*, and without opening the chest cavity, DURANT *et al.* [1954] recommended placing the body in the left lateral position to help displacement of air from the right ventricle. For arterial embolism, they recommended placing the body head downward. However, when access to the heart is made possible, massage of the heart is also widely recommended [ERICSSON *et al.*, 1964]. JONES and CROSS [1965] gave the following *methods* as being *useful to combat complication during cardiac surgery:* (1) filling the left heart with fluid; (2) insertion of a left arterial vent; (3) inducing ventricular fibrillation during the period that the heart is open; (4) holding the mitral valve incompetent with a special instrument, and (5) introduction of a vent into the left ventricle.

The use of a transvenous catheter for the aspiration of the entrapped air is also in common practice [TISOVEC and HAMILTON, 1967]. Correction of cerebral air embolism by regional perfusion with dextran is also recommended [ABBOTT, *et al.*, 1965].

V. Gas Embolism Due to Sudden Decompression

The blood flowing through the pulmonary capillaries is saturated with dissolved nitrogen to the PN_2 in the breathing mixture. Nitrogen is then carried to all the tissues of the body to saturate them. Since nitrogen is not metabolized by the body, it remains dissolved until the PN_2 in the lungs decreases, at which time the nitrogen is then removed from the body by the respiratory process. As a diver, breathing the air with 80% nitrogen, ascends from a dive, the elevated alveolar PN_2 falls. The dissolved nitrogen then diffuses from the tissues into the blood stream and from the blood stream into the lungs along the partial pressure gradient. If decompression is gradual, no harmful effects are observed. However, if the ascension is rapid, the dissolved nitrogen emerges from solution forming bubbles in the tissues and blood throughout the body. This phenomenon has been known as *decompression sickness*, the bends, caisson disease, diver's paralysis, and others.

A. Bubble Formation Due to Sudden Decompression

This section considers (1) the origin of gas bubbles which appear in organisms under various circumstances, namely nucleation; (2) stable nuclei; (3) criterion for cavitation, the condition for minute gas bubbles to grow, and (4) factors affecting their growth or disappearance. The works of HARVEY and his associates [HARVEY et al., 1944a, b] are good references for bubble formation in organisms.

1. Nucleation

Henry's law states that at equilibrium (i.e., saturation state) the dissolved gas concentration is proportional to the partial pressure of the gas in contact with the liquid, P_g. This pressure determines the gas tension in the liquid P_t. The difference between P_t and the hydrostatic pressure P_∞, ΔP ($=P_t-P_\infty$), is a measure of the tendency of gas to come of solution.

Two thoughts have been proposed to explain the mechanism of bubble formation in liquid, also named *cavitation:* homogeneous and heterogeneous nucleations. Homogeneous nucleation refers to the formation of bubbles in a homogeneous liquid at rest containing dissolved gas. The van der Waal equation theoretically estimates the amount of ΔP required for nucleation, to be of the order of 100 to 1,000 atm. There is little doubt that such a high ΔP arising from increase of gas tension alone cannot exist in the resting subject during exposure to a high altitude or in the usual pressure chamber. The theory of heterogeneous nucleation is based on the existence of a preformed gas phase or gas nuclei either suspended in the liquid or stuck to surfaces of container and foreign particles. Bubble formation is then meant by the growth of these gas nuclei to visible size with the application of an external distrubance such as ΔP. *In the human body, minute gas nuclei sticking to surfaces on the outside of cells grow and appear as bubbles in blood, lymph or intercellular fluids.*

After compression-decompression experiments with resting animals, bubbles are found abundantly throughout the vascular system in both arteries and veins (chiefly) and also in the fatty tissues.

2. Stable Nuclei

A nucleus is static or stable (meaning to exist indefinitely without change in size) when the gas pressure in the nucleus equals the gas tension in the liquid, i.e., $P_g = P_t$. This must be separately true for all gases involved. Under this situation, the pressure difference across the nucleus surface, $\Delta P = P_g - P_\infty$, is just balanced by the surface tension. Equation 37 gives for a spherical nucleus,

$$R_c = \frac{2\sigma}{\Delta P} \tag{41}$$

where R_c is called the critical radius. Equation 41 simply places a limit on the size of the nucleus: the largest diameter of the nucleus can be no greater than $2R_c$. Nuclei may be stable at or below this size. Let R be half the largest distance across the nucleus. Then, if a nucleus is stable, equation 41 dictates that the critical pressure difference $\Delta P_c \leq 2\sigma/R$. Nuclei will grow indefinitely if ΔP exceeds this critical value.

This relation is valid for both positive and negative values of P_∞. For a positive value of ΔP, equivalent to $P_g > P_\infty$, the nucleus is convex to the liquid, while for negative ΔP, R is also negative. A negative R means physically that a spherical nucleus has disappeared or the surface of an attached nucleus has been squeezed into a cavity and is concave to the liquid. Partly due to blood pressure, ΔP within the vascular system of a resting man at ground level varies from approximately zero in tissues to negative in arteries. Consequently, the gas mass on a vessel surface is nearly flat or has a curvature concave to the surface and is embedded in a depression or crack.

3. Criteria of Cavitation

On sudden decompression (referring to a sudden drop in P_∞), due to rapid ascent from a dive or to high altitude, $\Delta P = P_t - P_\infty$ abruptly changes to positive everywhere in the body. In small vessels ΔP remains locally high for a time due to CO_2 and N_2 of tissues and finally becomes zero to negative except for regions of fat deposit or very poor circulation.

If a minute gas phase is already present or formed under certain conditions, an increase in ΔP due to a sudden decompression may result in (i) a spherical nucleus to have R above the critical size in equation 41,

(ii) ΔP exceeding the critical pressure difference for nuclei attached to surfaces. This leads to the so-called cavitation, referring to the phenomenon in which gas nuclei are initiated to grow indefinitely. The growth is sustained by mass diffusion and will continue indefinitely, limited only by the amount of gas available or until the condition of $P_g = P_t$ is attained.

4. Factors Affecting Bubble Growth or Shrinkage

Diffusion varies as actual gas pressure in the nucleus minus gas tension in solution, gas solubility, gas diffusion coefficient and area of gas-liquid surface. Change in size of gas nuclei depends on (1) diffusion of gas into the nucleus, (2) creeping of the gas-liquid-solid boundary as determined by advancing and receding contact angles, (3) adjustment of the gas-liquid surface for minimum area consistent with the gas content of the nucleus and the hydrostatic pressure, and (4) thermodynamic behavior of the gas content in the nucleus. The third factor is affected by the inertia and viscosity of the liquid and is important when the liquid is under tension, i.e., P_∞ is negative. Quantitative treatment of bubble growth or collapse treated in section III.B.1.a) may be applied here to spherical nuclei.

B. Symptoms of Decompression Sickness

Air embolism associated with deep sea diving was first reported by BERT in 1878, one century ago. He attributed 'the bends' to the liberation of nitrogen that has been dissolved in the body under high pressure following a sudden decompression. These gas bubbles have since been observed in the blood and tissues. Formation of bubbles is more favorable on decompression from sea level pressure to that of higher altitude than from higher pressures beneath the sea to the surface.

Symptoms of decompression sickness have occurred with the following frequencies: discomfort and local pain, particularly in the limbs, 89%; dizziness, 5.3%; paralysis, 2.3%; shortness of breath, 1.6%; extreme fatigue and pain, 1.3%; and collapse with unconsciousness, 0.5%. These symptoms usually appear within a few minutes to an hour after sudden decompression but occasionally develop six or more hours after decompression. The most common symptoms are discomfort and pain. They are attributed to the distortion, extension, and rupture of the tissue fibers due to the formation and expansion of gas bubbles causing hemorrhages and local ischemia. Sometimes the pain is caused by bubble formation in pain-sensitive tissues such as the joints and bones. It can also result from sudden distension of the gastrointestinal gases causing severe bloating of the gut. The most serious, although less frequent, cases are bubble formation in the nervous systems. Mechanical rupture of fiber pathways may occur which leads to serious mental disorders or permanent paralysis. The shortness of breath or 'the chokes' is due to the blockage of pulmonary blood flow by the bubbles which are formed in the blood and become caught in the capillaries of the lungs.

Muscular activity favors bubble formation following decompression. This is attributed to high local concentration of carbon dioxide formed during muscular activity. The larger and heavier is the individual, the greater is the tendency.

C. Prevention and Treatment of Decompression Sickness

If decompression is gradual, the dissolved nitrogen can be eliminated through the lungs to prevent decompression sickness. Figure 14 shows the rate of nitrogen release from the body when a diver has come to sea level from prolonged exposure to compressed air at 10 m depth. The liberation of nitrogen from the water of the body is completed in about 1 h. During this period, about two-thirds of nitrogen is released from the fat of the body. Approximately 90% of the total nitrogen is liberated in 6 h. However, the diver is not completely safe so long as some excess nitrogen still remains in the body. Therefore, a diver who has been beneath the sea for a long time must be decompressed for many hours. A decompression table compiled by the US Navy may be used as a guidance in determining the total decompression time and the optimal time on bottom.

Oxygen may be administered to a diver for more rapid decompression as he ascends closer to the surface of the sea (to tolerate higher than normal oxygen partial pressures). The nitrogen partial pressure in his alveoli is then considerably reduced resulting in an increase in the rate of

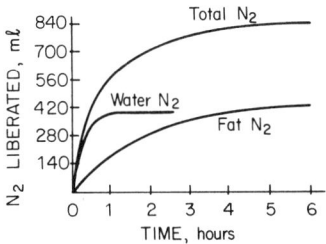

Fig. 14. Rate of nitrogen release following sudden decompression from 10 m beneath the ocean.

nitrogen removal from his body fluids. The diver can then be brought to the surface more rapidly. *A second way to prevent decompression sickness* is to place the diver in a decompression tank within 5 min after he arrives at the surface. An appropriate decompression table is used to prevent bubble formation. *A third means of prevention* is to breathe helium-oxygen mixtures in deep dives.

Helium has advantages over nitrogen in deep dives, including (1) a shorter decompression time, (2) no narcotic effect, and (3) lower airway resistance in the lungs. Decompression time is reduced because of (1) less dissolution of helium in the body, about 40% that of nitrogen, and (2) higher diffusion coefficient of helium through the tissues, about 2.5 times that of nitrogen, due to its smaller atomic size. However, helium is inferior to nitrogen for short, shallow dives resulting from two reasons: (1) lower supersaturation of helium – cavitation begins when the PHe in the body fluids is only 1.7 times the pressure on the outside of the body compared with 3.0 for nitrogen. (2) Rapid diffusion of helium – more helium than nitrogen is dissolved in the body fluids in a short time.

The *treatment of decompression sickness* involves: (1) increasing the ambient pressure in order to reduce bubble size. This procedure is called recompression and can be accomplished in a decompression chamber. (2) Breathing oxygen in order to increase the gradient for nitrogen to be liberated from the body and to increase tissue oxygenation. Under certain circumstances, such as very high ambient pressures, oxygen-inert gas mixtures must be breathed in order to avoid oxygen toxicity.

The US Navy Diving Manual [1972] advocates breathing helium-oxygen mixtures. However, experiments [STRAUSS and KUNKLE, 1974] and theory [YANG, and LIANG, 1972] have revealed that nitrogen bubbles are expected to grow, since helium would diffuse into them faster than nitrogen would diffuse out. Consequently, decompression sickness could be worsened by switching from nitrogen to helium in the breathing gas mixture. BAUER *et al.* [1965] recommended *hypothermic treatment* of 'chokes'. This suggestion is sound *since gas bubbles shrink in volume when heat is removed from them* and the surrounding body fluids or tissues due to analogy between heat and mass transfer (The latter case is mentioned earlier in section III.B.1.a.).

VI. Tissue-Capillary Gas Exchange

Transfer of oxygen from the atmospheric air to the cells of the body takes three steps: first is breathing which inhales the gas for the oxygenation of the blood in the lungs, followed by its carriage to the tissues, and finally migration of oxygen from the blood to the cells. Carbon dioxide, one waste product from cellular activities, is transferred from the cells to the atmosphere following the same pathways by oxygen but in the opposite direction. The third stage of oxygen-carbon dioxide exchange between the blood stream and the cells is called the *internal exchange.*

Sometimes, a gas mass is purposely introduced (by injection) into subcutaneous tissues, liver and perirenal fat. The subcutaneous gas pocket, an *in vivo* tonometry system, has been proved an excellent tool in obtaining basic information concerning tissue-capillary gas exchange.

A. Exchanges of Oxygen and Carbon Dioxide in the Tissues

The rate at which the tissue cells of the body consume oxygen is equal to the rate of its intake by the lungs. Similarly, the rate at which the cells produce carbon dioxide is equal to the rate of its discharge from the lungs. The balance between them is always preserved even when these rates are accelerated during strenuous exercise. The internal exchange between the blood stream and the cells takes place in two steps: oxygen migrates from the blood to the tissue fluid surrounding the cells and then from the tissue fluid into the cell interior, while carbon dioxide is transferred in the opposite direction with the tissue fluid as acting as an intermediary.

1. Exchange of Oxygen

Arterial blood that is freshly oxygenated in the lungs reaches the tissue capillaries at a PO_2 of about 96 mm Hg. It corresponds to about 97.5% hemoglobin saturation and 19.5 ml oxygen in each 100 ml of blood, as indicated by point A in the oxyhemoglobin dissociation curve (fig. 4). The PO_2 within the interior of tissue cells is approximately 35 mm Hg. Like the uptake of oxygen in the lungs, *the PO_2 difference causes passive diffusion of dissolved oxygen from the plasma through the capillary walls to reach the cells.* With oxygen lost from the plasma, the PO_2 surrounding the red corpuscles goes down, causing oxyhemoglobin to dissociate and release about one-quarter of its available oxygen for tissue consumption. As a result, the PO_2 in the venous blood from the tissue capillaries is about 40 mm Hg, corresponding to about 75% saturated with hemoglobin and an oxygen load of approximately 14.5 ml per 100 ml blood as given by point VR (fig. 4). The arteriovenous oxygen difference is 5 ml for each 100 ml of blood. This means that the cells consume 250 ml of oxygen per minute in a resting man whose blood circulation is 5,000 ml in 1 min. However, during vigorous exercise, the arteriovenous oxygen difference may reach 15 ml per 100 ml blood as marked by point VE in the shifted dissociation curve. Meanwhile the total blood flow circulating the body may reach 30,000 ml/min, resulting in 4,500 ml of oxygen supply to the tissues and organs each minute.

The length of the tissue capillaries varies from 0.4 to 0.7 mm which takes 0.5–1.0 sec for blood to traverse in a resting man.

2. Exchange of Carbon Dioxide

Carbon dioxide is formed in the cells through metabolism resulting in a PCO_2 of about 46 mm Hg. The PCO_2 of arterial blood reaching the tissues is approximately 40 mm Hg. Like the interchange of oxygen, this PCO_2 difference causes dissolved carbon dioxide molecules to diffuse through the tissue fluid into the plasma of the capillary blood, finally entering the red corpuscles. The diffusion process of carbon dioxide is much more rapid than that of oxygen because of high solubility and diffusion coefficient of the gas.

The hydration of CO_2 (eq. 8) in the plasma and tissue fluid is a slow reaction, and the dissociation of carbonic acid (eq. 7) is slight. On the other hand, these two reactions in the red corpuscles are very fast due to the presence of the enzyme catalyst. The products of dissociation, H^+ and CO_3^-, do not accumulate within the red corpuscles to interfere with the reactions: the hydrogen ion is accepted by hemoglobin to form reduced hemoglobin (eq. 9), while the carbonate ions diffuse outwards into the plasma and tissue fluid from which chloride ions and other anions diffuse into the red corpuscles. Carbon dioxide also reacts directly with hemoglobin to form carbaminohemoglobin, an extremely rapid reversible reaction requiring no special enzyme.

B. Dissolution of Gas Emboli in the Tissues

When pure oxygen is breathed, CO_2 and O_2 are the only gases presented in the circulating blood. If then a subcutaneous gas pocket with a foreign gas is introduced, the gas will migrate through the tissues and finally into the blood stream. Any foreign gas absorbed by the blood will eventually be eliminated in the lungs. Therefore, the arterial blood which bathes the subcutaneous pocket has a foreign gas pressure of zero.

Like the exchange of CO_2, that of inert gases across the gas pocket wall is a passive diffusion caused by the gas pressure difference between the gas pocket and the blood stream. According to Fick's first law of mass diffusion, the volumetric absorption rate of the inert gas m by blood $\dot{V}_m$ is

$$\dot{V}_m = (D\alpha_t)_m A \frac{(P_g - \bar{P})_m}{l} \tag{42}$$

where D_m (cm²/sec) and α_{tm} (ml gas/ml blood/atm) are the diffusion and solubility coefficients of the gas in the diffusion barrier, respectively; A, the area for diffusion; P_{gm}, the gas pressure in the pocket; $\bar{P}_m$, the mean gas pressure in the blood; l, the thickness of the diffusion barrier; and m, the subscript indicating the inert gas m. The product $D\alpha$ is termed 'permeation coefficient'.

The uptake of the inert gas by blood perfusing the gas pocket area $\dot{V}_m$ can be expressed as

$$\dot{V}_m = \dot{Q} \cdot \alpha_{bm}(P_2 - P_1)_m. \tag{43}$$

Here, $\dot{Q}$ is the blood (solvent) flow rate (ml blood/sec), α_{bm} is the solubility coefficient of the inert gas m in blood, and P_{1m} and P_{2m} are the blood pressures entering and leaving the gas pocket region, respectively. $\bar{P}_m$ in equation 42 is the arithmetical average of P_{1m} and P_{2m}. The conservation of mass requires that these two $\dot{V}_m$ values in equations 42 and 43 must be equal. Consequently, it yields

$$\frac{D_m A}{l\dot{Q}} = \left(\frac{\alpha_b}{\alpha_t}\right)_m \left(\frac{P_2 - P_1}{P_g - \bar{P}}\right)_m \cong \left(\frac{P_2 - P_1}{P_g - \bar{P}}\right)_m.$$

Since the solubility coefficients α_{tm} and α_{bm} are about equal, one gets

$$H \cong \left(\frac{P_2 - P_1}{P_g - \bar{P}}\right)_m \tag{44}$$

where H is defined as $D_m A/l\dot{Q}$, ml tissue/ml blood. A large value of H signifies that the pressure change of the inert gas m in the blood perfusing the pocket is larger than the difference between the gas pressure inside the pocket and the mean pressure of the gas dissolved in the blood. The latter is the driving potential for mass transfer of the gas from the pocket to the blood.

Quantitative data obtained with the subcutaneous gas pocket provide basic information concerning tissue-capillary gas exchange: specifically for the determinations of (i) permeation of the dissolved gases in the tissues and their interaction with hemoglobin; (ii) tension of the dissolved gases in the tissues in the constant composition state, and (iii) the influx and efflux of inert gases across the surface of a decomposition bubble.

1. In vivo Tests

The experimental study of tissue-capillary gas exchange using a subcutaneous gas pocket in rats is generally performed in two stages:

preparatory and test stages [PIIPER *et al.*, 1962; VAN LIEW, and PASSKE, 1967].

In the preparatory stage, air-filled gas pockets of about 20–30 ml are maintained for several days (from 5 to 30 days depending upon investigators) during which time the local tissue reaction subsides. By the 5th day, the measured PO_2 and PCO_2 of pocket gases reach reasonably constant values, indicating that blood perfusion to the pocket has stabilized. For the following 3 weeks such pockets are in a state of constant composition of oxygen and carbon dioxide and are satisfactory for use to study the disappearance of test gases.

In the test stage, 20 ml of test gas are injected into the pocket after evacuation of the preparatory air. In about 2 h after the introduction of the foreign gas, the oxygen and carbon dioxide composition of the pocket will be constant and remain so until all the gas has been absorbed. The subsequent time course of the volume of this pocket is then determined by extracting the gas from the pocket into a calibrated syringe, noting the volume and time, and then reinjecting the gas from the syringe back into the pocket. The subcutaneous gas pockets containing the foreign gas – air, inert gases such as argon, helium, nitrogen, hydrogen, cyclopropane (C_3H_6), nitrous oxide (N_2O) and sulfur hexafluoride (SF_6), or reacting gases such as oxygen, carbon monoxide and carbon dioxide – have been tested in airbreathing, oxygen breathing or dead rats.

PIIPER *et al.* [1962] have estimated the order of magnitude of the thickness of the diffusion barrier as 10–100 μm. Table IV illustrates the factor $(D\alpha)_t A/l$ for permeation of various gases from subcutaneous pockets, obtained by dividing the exit rate of the test gas by its partial pressure difference between the inside and outside of the pocket.

If a pocket contains a slower-diffusing gas than the gas which is breathed, a transient increase of pocket size can be expected. The

Table IV. The factor ($D\alpha A/l$) for permeation of various gases from non-fatty subcutaneous gas pockets in rats

Gas	$D\alpha A/l$ ml/h/mm Hg	Reference
Argon (A)	19	PIIPER *et al.*, 1962
Hydrogen (H_2)	17	PIIPER *et al.*, 1962
Helium (He)	12	PIIPER *et al.*, 1962
Sulfur hexafluoride (SF_6)	5	PIIPER *et al.*, 1962
Nitrogen (N_2)	10	PIIPER *et al.*, 1962
Nitrogen (N_2)	11	VAN LIEW, and PASSKE, 1967
Neon (Ne)	8	VAN LIEW, and PASSKE, 1967

situation is observed in the SF_6 bubble as shown in figure 15. On the other hand, if the pocket contains a gas which diffuses faster than the breathing gas, a transient decrease of bubble size is seen in figures 16–20. These results can be applied to a change of breathing mixture during a decompression. For example, a man who breathed N_2 during the dive is suffering from N_2 bubbles following a sudden decompression. If he switches to a breathing mixture containing a slightly faster gas like helium, the bubbles can grow because helium will enter the bubbles more rapidly than the nitrogen will escape. As a result, the decompression symptoms will increase [STRAUSS and KUNKLE, 1974].

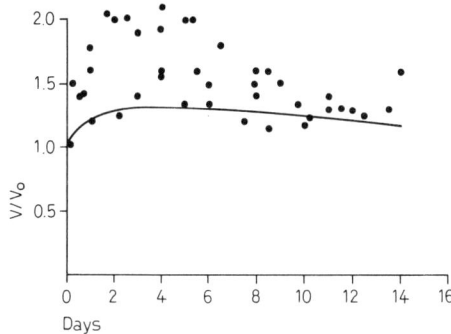

Fig. 15. Time course of sulfur hexafluoride gas in a subcutaneous pocket of an air-breathing rat.

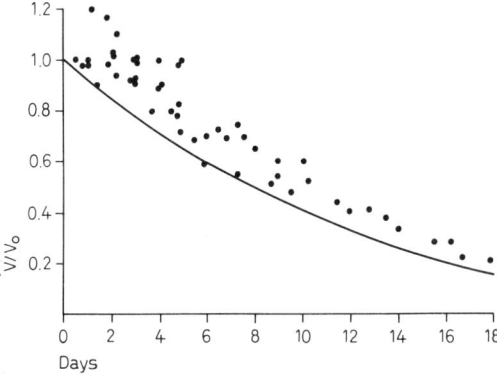

Fig. 16. Time course of nitrogen gas in a subcutaneous pocket of an air-breathing rat.

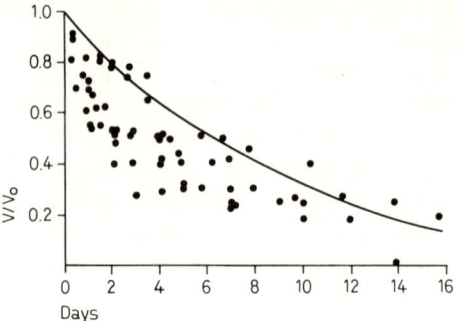

Fig. 17. Time course of helium gas in a subcutaneous pocket of an air-breathing rat.

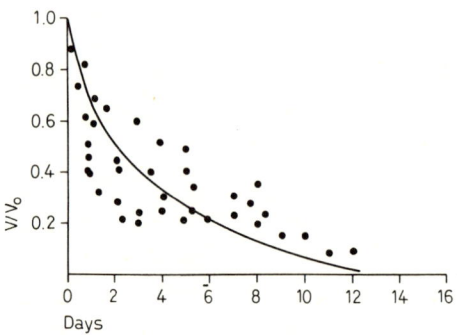

Fig. 18. Time course of argon gas in a subcutaneous pocket of an air-breathing rat.

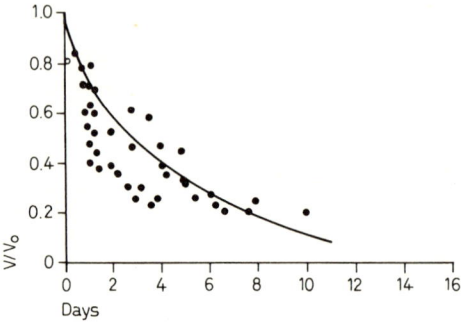

Fig. 19. Time course of hydrogen gas in a subcutaneous pocket of an air-breathing rat.

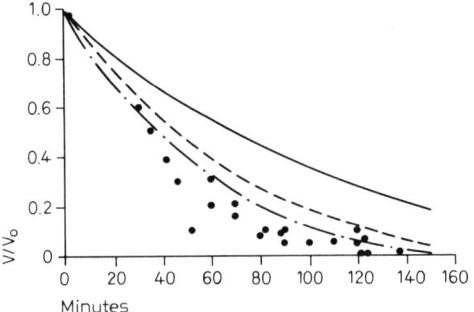

Fig. 20. Time course of nitrous oxide gas in a subcutaneous pocket of an air-breathing rat (from top: $\alpha \times 10^5 = 68.1$; 68.1; 120; $(v \times 10 = 4.35$; 4.35; 4.35; $kQ = 0.214$; 0.50; 0.214).

2. Theory

Depending on the physical characteristics of the foreign gas and the breathing environment, the migration of a single gas or two gases may occur across the pocket-tissue interface. The single-gas case is characterized by the influx of any gas in an oxygen-breathing animal, or of nitrogen in an air-breathing animal. If, instead of nitrogen, some other gas (inert or reacting gas including air) is injected into the pocket, and if the animal breathes atmospheric air, the pocket will soon contain two gases, nitrogen from the blood and the foreign gas injected. This corresponds to the two-gas case in which the fluxes of the gases must be reckoned with. These two fluxes are interdependent, that is, the efflux of either one alters the driving pressure gradient of the other. If the foreign inert gas such as SF_6 is nearly insoluble, the pocket volume will increase substantially over initial volume before it begins to decrease toward zero. On the other hand, a pocket of highly soluble inert gas such as N_2O will decrease rapidly, even before more than a small quantity of nitrogen can enter.

The *dynamics of a gas cavity* in a tissue-capillary system may be considered as a problem of determining the distribution of the dissolved-gas tension under the action of the difference in pressure in the tissue at the interface and at the point at a large distance from the gas cavity. The gas tension in the tissue at the interface depends on the pressure exerted by the gas in the cavity. Any external force variation may be imposed on the tissue through muscular activity. However, it will induce a creep process in the system which, in turn, affects the gas pressure of the cavity and consequently its difference with the gas tension in the tissue at a large distance from the cavity. It is the pressure difference which will induce the migration of the dissolved gas to or from the cavity. Hence, there is a coupling between the equation of motion (or stress equilibrium equation) and the equation for the pressure (or concentration) field in the tissue-capillary system.

PIIPER et al. [1962] have analyzed the gas exchange between a subcutaneous gas pocket and the tissue-capillary system for the single-gas diffusion case by treating the tissue-capillary system as a lumped-parameter unit. Using the magnitude of the parameter H as a criterion, they have divided the gas absorption process into *three categories:* (i) perfusion controlling mechanism for very large values of H; (ii) diffusion controlling mechanism for very small values of H, and (iii) perfusion-diffusion controlling mechanism for intermediate values of H. The work was extended to the two-gas diffusion case by TUCKER and TENNEY [1966]. VAN LIEW [1968] has given a theoretical analysis for both the single- and two-gas diffusion cases by treating the tissue-capillary system as a distributed-parameter unit.

Theoretical approach can be proceeded under three categories: (i) tissue creep controlled; (ii) mass transfer controlled, and (iii) the intermediate case where both mechanisms are of comparable importance. Case (i) fits a man under muscular activity, while case (ii) applies to a resting subject with the gas pocket in a state of constant composition. Thus far, only case (ii) has experienced an extensive application in an *in vivo* tonometry of gas permeation in the tissues.

Consider a spherical gas cavity of radius R_0 (mean radius if nonspherical) situated in a homogeneous and isotropic tissue-capillary system with the dissolved gas of uniform tension P_∞. Initially, the total gas pressure in the cavity is P_{g0}, which is in equilibrium with the surface tension $2\sigma/R_0$ and the uniform shear stress in the system $\tau_\infty(0)$. An external load $\tau_\infty(t)$ is then impressed on the system at a large distance from the cavity. At any time $t > 0$, the gas pressure in the cavity is P_g and the dissolved-gas tension and the shear strain in the tissue-capillary system are P and ε_{rr}, respectively. Due to the variations in the pressure $(P_g - P_\infty)$ and strain rate $\dot{\varepsilon}_{rr}$, the cavity may grow or shrink depending upon the nature of the gas, of the tissue-capillary system, and of the stress applied on the cavity wall. The growth or collapse of the gas cavity is accompanied by the migration of the gas across the cavity wall. Both the gas cavity and the tissue-capillary system are isothermal at the same temperature T.

a) Tissue Creep Controlling

The stress-strain relation of the tissue in creep is described by the standard linear model of viscoelasticity, in a spherical polar coordinate system (r, θ, ϕ), as

$$(1+\lambda D)(\tau_{rr}-\tau_{\theta\theta}) = \eta(1+\mu D)\epsilon_{rr}. \qquad (45)$$

Here, τ_{rr} and $\tau_{\theta\theta}$ are the normal components of the shear stress in the r and θ direction, respectively, D is the differential operator, and λ, η and μ denote the rheological constants of the tissue-capillary system. Following the intepration of continuity and equilibrium equations, and the equation defining the creep rate (subject to the appropriate initial and boundary conditions, we obtain the dynamic equation of a gas cavity in tissues as

$$\tau_\infty + P_g - \frac{2\sigma}{R} + \lambda\left(\dot{\tau}_\infty + \dot{P}_g - \frac{2\sigma}{R^2}\dot{R}\right) + \Delta\tau_{rr} = \frac{4\eta}{9R^3}(R^3 - R_0^3 + 3\mu R^2 \dot{R}) \qquad (46)$$

subject to the initial conditions

$$R(0) = R_0, \dot{R}(0) = 0, \tau_\infty(0) = -P_{g0} + \frac{2\sigma}{R_0}, \quad \dot{\tau}_\infty(0) = 0, P_g(0) = P_{g0}, \dot{P}_g(0) = 0 \qquad (47)$$

where

$$\Delta\tau_{rr0} = -\tau_{rr}(R_0, 0) + \tau_{rr}(\infty, 0)$$

and τ_∞ is identical with $\tau_{rr}(\infty, t)$. The gas in the cavity is assumed to undergo a reversible polytropic process; the variations of its pressure and volume can be related as

$$P_g(t) = P_{g0}(R/R_0)^{3n} \qquad (48)$$

Here n is the polytropic exponent and takes the value of unity for an isothermal process.

The instantaneous rate of the gas escaping from the pocket to the tissue-capillary system can be determined by equation 44 in which P_g is from equation 48 and $\bar{P}$ corresponds to P_∞.

b) Mass Transfer Controlling

In the diffusion of a gas out of decompression bubbles or artificially injected gas pockets, the gas diffusing through living tissues meets blood capillaries at various depths. Either the solution of the gas in blood or its chemical reaction with blood elements has created multiple small sinks for the gas within the diffusion barrier. Thus, the dissolved gas concentration decreases with distance from the gas-tissue interface because of two factors: (i) blood leaving capillaries removes the gas from the system-mass consumption (i.e., blood perfusion) and (ii) divergence of radii of the sphere of tissue around the pocket-mass diffusion. The difference between the rates of mass diffusion and mass consumption should be equal to the time-rate change in the amount of locally dissolved gas according to the principle of mass conservation.

(i) *Inert gases.* For inert gases such as nitrogen, argon, helium, hydrogen, nitrous oxide and sulfur hexafluoride which are not metabolized and which do not enter into chemical reaction with blood elements, disappearance of gas from the diffusion system will be by solution in blood. The rate of disappearance is directly proportional to the difference between the dissolved gas pressure P and that in the arterial blood, P_∞. Thus, the diffusion equation reads

$$\frac{\partial P}{\partial t} = \frac{D}{r^2} \frac{\partial}{\partial r}\left(r^2 \frac{\partial P}{\partial r}\right) - E(P - P_\infty) \tag{49}$$

where $E = \alpha_b kQ/\alpha_t$; α_b, the gas solubility in blood; α_t, the gas solubility in tissue; k, the coefficient of end capillary saturation; and Q, the rate of actual blood perfusion. kQ is the effective blood perfusion. The appropriate initial and boundary conditions are

$$P(r, 0) = P_\infty$$
$$P(R, t) = P_g, \quad P(\infty, t) = P_\infty. \tag{50}$$

The pressure gradient at $r = R$ is obtained by solving equation 49 subject to equation 50 as [YANG et al., 1971]

$$\left(\frac{\partial P}{\partial r}\right)_{r=R} = (P_g - P_\infty) I_1(t) \tag{51}$$

where

$$I_1(t) = \frac{1}{R} + \frac{\exp(-Et)}{(\pi Dt)^{1/2}} + (E/D)^{1/2} \operatorname{erf}(Et)^{1/2}.$$

At large times, $I_1(t)$ approaches the value

$$I_1(t) \cong \frac{1}{R} + \left(\frac{E}{D}\right)^{1/2}.$$

(ii) Reacting gases. The reacting gases include oxygen, carbon dioxide and carbon monoxide. The diffusion and perfusion of carbon dioxide will not be considered here due to complexity resulting from chemical combination with blood elements, blood perfusion or metabolism. Consideration is given only to the following three cases: (α) oxygen in dead animal – the gas diffuses out and disappears due to utilization by the tissue; (β) carbon monoxide – the gas diffuses out and disappears by combining with hemoglobin in the blood perfusing in the tissue; and (γ) oxygen in living animal – the gas disappears both by metabolism and by carriage in blood. The *diffusion equation* applicable to all these three cases may be expressed as

$$\frac{\partial P}{\partial t} = \frac{D}{r^2} \frac{\partial}{\partial r}\left(r^2 \frac{\partial P}{\partial r}\right) - B \tag{53}$$

where $B = q/\alpha_t$ for case (α) and $B = Q \Delta C/\alpha_b$ for cases (β) and (γ). Here, q is the rate of oxygen consumption in the tissue and ΔC is the change of the dissolved gas content in the blood. The appropriate initial and boundary conditions are identical with equation 50, in which $P_\infty = 0$ in cases (α) and (β) and $P_\infty = 40$ mm Hg in case (γ).

The pressure gradient at $r = R$ is obtained as

$$\left(\frac{\partial P}{\partial r}\right)_{r=R} (P_\infty - P_g) I_2(t) \tag{54}$$

where

$$I_2(t) = \frac{1}{R} + \frac{1}{(\pi D t)^{1/2}} + B^* t \left(\frac{1}{R} + \frac{2}{(\pi D t)^{1/2}}\right) \tag{55}$$

and $B^* = B/(P_g - P_\infty)$. At large times, I_2 approaches

$$I_2(t) = (1 + B^* t) R.$$

Problems of this kind can be further classified into two categories, i.e., (i) single-gas diffusion and (ii) multi-gas diffusion.

(i) *Single-gas diffusion case.* For a pocket of nitrogen in an air-breathing animal or for any inert gas in an animal breathing oxygen, the volume change of the pocket is determined by single gas diffusion. If the gas inside the pocket behaves like an ideal gas, its equation of state is

$$\frac{4\pi R^3 P_g}{3} = m\bar{R}T \tag{56}$$

where $\bar{R}$ is the gas constant.

The rate of mass diffusion $\dot{m}$ can be related to the pressure gradient in the tissue at the pocket wall (by Fick's law) by the expression

$$\dot{m} = 4\pi R^2 D H \left(\frac{\partial P}{\partial r}\right)_{r=R} \tag{57}$$

where $H = \alpha_t/\rho_g$ and ρ_g is the gas density in the pocket. Now, equation 56 is differentiated with respect to time for small variations of P_g. The resulting expression is combined with equations 51 and 57 to yield *the bubble dynamic equation for the single-gas diffusion case:*

$$P_g \dot{R} = D H \bar{R} T (P_\infty - P_g) I_i \quad (i = 1, 2). \tag{58}$$

The appropriate initial condition is $R(0) = R_0$. The volume history of the pocket may be obtained by the simultaneous numerical integration of equations 58, 52 and 55.

(ii) *Multi-gas diffusion case.* If instead of nitrogen, some inert gas such as argon, hydrogen, helium, sulfur hexafluoride, cyclopropane or nitrous oxide is initially injected into the pocket, and if the animal breathes normal atmospheric air, two gases will cross the pocket wall, nitrogen from the blood and the foreign inert gas injected. In that case, the relative rate of movement of the two gases determines the volume history of the gas pocket. If each gas behaves ideally in the pocket, then each gas component satisfies the equation of state (eq. 56) which is rewritten for the j-th gas components as

$$P_{gj} = \frac{3m_j \bar{R}_j T}{4\pi R^3}. \tag{59}$$

The rate of mass diffusion $\dot{m}_j$ of the j-th gas component across the pocket wall is

$$\dot{m}_j = 4\pi R^2 D_j H_j \left(\frac{\partial P_j}{\partial r}\right)_{r=R}. \tag{60}$$

With the aid of Dalton's law

$$P_g = \sum P_{gj} \tag{61}$$

and, following the procedure for the single-gas diffusion case, one obtains *the bubble dynamic equation for the multi-gas diffusion case* as

$$P_g \dot{R} = T \Sigma D_j H_j \bar{R}_j (P_\infty - P_g)_j I_j. \tag{62}$$

It should be noted that since diffusion of nitrogen and the inert gas is taking place simultaneously, the partial pressure of each gas, P_{gj} in equation 59, varies with time. P_{gj} can be determined, however, by integrating $\dot{m}_j$ in equation 60 with respect to time from zero to t and by substituting m_j thus obtained into equation 59. The expression is found to be

$$P_{gj} = \frac{3\bar{R}_j T}{4\pi R^3} \left[4\pi D_j H_j \int_0^t R^2 \left(\frac{\partial P_j}{\partial r}\right)_{r=R} dt + m_{0j} \right] \tag{63}$$

where m_{0j} denotes the initial mass of the j-th gas in the pocket. Numerical results can be obtained by solving equations 63, 62, 52 and 55 simultaneously with the use of a digital computer.

c) Both Tissue Creep and Mass Transfer Controlling

This is the general case in which both the creep and mass transfer processes in the tissue-capillary system affect the growth and shrinkage of the gas cavity. The volume history of the cavity can be obtained by simultaneously solving equations 58 or 62 with equations 52 and 55, subject to the initial conditions 47 using a digital computer.

Table V. Physical and physiological properties for some gases ($\alpha_b = \alpha = \alpha_t$)

Gas	Solubility, $\alpha \times 10^5$ ml·ml^{-1} blood ·mm Hg^{-1}	Permeation coefficient $D\alpha \times 10^8$ cm^2·min^{-1}·mm Hg^{-1}	Effective perfusion rate kQ min^{-1}
N_2	1.84	1.00	0.26
A	4.1	1.87	0.223
H_2	2.12	1.94	0.33
He	1.25	1.44	0.33
SF_6	0.606	0.145	0.13
N_2O	68.0	29.6	0.214

3. Comparison of Theory with in vivo Tests

Theoretical results are available for the mass transfer controlling case – the inert gas exchange in subcutaneous gas pockets of air-breathing animals using the physical and physiological data listed in table V [YANG and LIANG, 1972]. In the table, Q is 0.33 mm of blood/min/cm^3 of tissue. The value of k, the coefficient of end-capillary saturation, is taken from figure 12 of VAN LIEW [1968]. The gas solubility in blood, α_b, is equal to that in tissue, α_t. These results are presented in figures 15–20 for comparison with the experimental data of TUCKER and TENNEY [1966]. The ordinate represents the relative volume (the ratio of the instantaneous volume V to the initial volume V_0). The solid lines are theoretical curves obtained by plotting equation 58 in the case of nitrogen or (eq. 62) in the cases of other inert gases. The *in vivo* test data are indicated as solid dots. The total pressure inside the gas pocket is taken to be 1 atm in the calculation. The contribution of fluxes of O_2, CO_2 and H_2O is neglected since these three gases occupy only 11% of the total volume [TUCKER and TENNEY, 1966]. Several different inert gas pockets are compared in the figures: a nitrogen gas pocket exemplifies the single-gas diffusion case. SF_6 gas is nearly insoluble and represents the diffusion-controlling exchange process. On the other hand, N_2O gas is highly soluble and represents the perfusion-controlling exchange process. The exchange process for argon, hydrogen or helium is both diffusion- and blood-flow-determined. The validity of the mathematical model is borne out in the *agreement between theory and experiment for all cases*.

References

ABBOTT, O. A.; EXARKOS, N., AYDIN, K.: Immediate correction of cerebral air embolism by regional perfusion with dextran: experimental and clinical observations. 1st Annu. Meet., Soc. Thoracic Surgery, St. Louis, Mo. 1965.

BAUER, R. O.; CAMPBELL, S.; GOODMAN, R.; MUNSAT, T. L., and POPS, M. A.: Aeroembolism treated by hypothermia. Aerospace Med. *36:* 671–675 (1965).

BERNSTEIN, E. F.; BLACKSHEAR, P. L., jr., and KELLER, K. H.: Factors influencing erythrocyte destruction in artificial organs. Am. J. Surg. *114:* 126–138 (1967).

BURBANK, A.; FERGUSON, T. B., and BURFORD, T. H.: Carbon dioxide flooding of the chest in open heart surgery. J. thorac. cardiovasc. Surg. *50:* 691–698 (1965).

CHAN, K. S. and YANG, WEN-JEI: Behavior of gas emboli subjected to pressure variation in biological systems. J. Biomech. *2:* 151–156 (1969a).

CHAN, K. S. and YANG, WEN-JEI: Survey of literature pertinent to the problems of gas embolism in human body. J. Biomech. *2:* 299–312 (1969b).

DURANT, T. M.; OPPENHEIMER, M. J.; LYNCH, P. R.; OSCANIO, G., and WEBBER, D.: Body position in relation to venous air embolism: a roentgenologic study. Am. J. med. Sci. *227:* 509–520 (1954).

EGUCHI, S.; SAHURAI, Y., and YAMAGUCHI, A.: The use of CO_2 gas to prevent air embolism during open heart surgery. Acta med. biol., Niigata *11:* 1–13 (1963).

ERICSSON, J. A.; GOTTLIEB, J. D., and SWEET, R. B.: Closed chest cardiac massage in the treatment of venous air embolism. New Engl. Med. *270:* 1353–1354 (1964).

FORSTER, R. E.: Rate of gas uptake by red cells; in FENN and RAHN, Handbook of physiology, sect. 3: Respiration, chapt. 32 (1964).

FORSTER, R. E. and VAN DE LINDT, W. J.: Calculations of the rates of uptake of O_2 and CO by red cells using a digital computer. Fed. Proc. *18:* 47 (1959).

FUNG, Y. C.: Theoretical considerations of the elasticity of red cells and small blood vessels. Fed. Proc. *25:* 1761–1772 (1966).

GALLETTI, P. M. and BRECHER, G. A.: Heart-lung bypass. Principles and techniques of extracorporeal circulation, chapt. 17 (Grune & Stratton, New York 1962).

GIBSON, Q. H.: The kinetics of reactions between hemoglobin and gases. Prog. biophys. Chem. *9:* 1–53. (1959).

GOTTLIEB, J. D.; ERICSSON, J. A., and SWEET, R. B.: (1965): Venous air embolism. Anesth. Analg. curr. Res. *44:* 773–779 (1965).

HARVEY, E. N.; BARNES, D. K.; MCELROY, W. D.; WHITELEY, A. H.; PEASE, D. C., and COOPER, K. W.: Bubble formation in animals. I. Physical factors. J. cell. comp. Physiol. *24:* 1–22 (1944a).

HARVEY, E. N.; WHITELEY, A. H.; MCELROY, W. D.; PEASE, D. C., and BARNES, D. K.: Bubble formation in animals. II. Gas nuclei and their distribution in blood and tissues. J. cell. comp. Physiol. *24:* 23–34 (1944b).

JONES, R. D. and CROSS, F. S.: A vent valve to minimize air embolism during open-heart surgery. J. thorac. cardiovasc. Surg. *48:* 310–313 (1965).

KLUG, A.; KREUZER, F., and ROUGHTON, F. J. W.: Simultaneous diffusion and chemical reaction in thin layers of hemoglobin solution. Proc. R. Soc. B *145:* 452–472 (1956).

LUBBERS, J.; TENHOF, J. P.; VAN DER VEEN, P. H.; VAN DER BERG, J. W.; DORLAS, J. C., and VAN DER HEIDE, J. N. H.: Microgasemboli in a pump oxygenator during open-heart surgery. Archvm chir. neerl. *26:* 40–53 (1974).

MARCHAND, P.; HASSELT, H. VAN, and LUNTZ, C. H.: Massive venous air embolism. S. Afr. med. J. *38:* 202–208. (1964).

MUNSON, E. S. and MERRICK, H. C.: Effect of nitrous oxide on venous air embolism. Anesthesiology *27:* 783–787 (1966).

NICOLSON, P. and ROUGHTON, F. J. W.: A theoretical study of the influence of diffusion and chemical reaction velocity on the rate of exchange of carbon monoxide and oxygen between the red blood corpuscle and surrounding fluid. Proc. R. Soc. B *138:* 241–264 (1951).

PIERCE, E. C., II: Extracorporeal circulation for open-heart surgery, chapt. 2 (Thomas, Springfield (1969).
PIIPER, J.; CANFIELD, R. E., and RAHN, H.: Absorption of various inert gases from subcutaneous gas pockets in rats. J. appl. Physiol. *17*: 268–274 (1962).
SCHLICHTING, H.: Boundary layer theory; 2nd ed., pp. 607–610 (McGraw-Hill, New York 1960).
SELMAN, M. W.; MCALPINE, W. A.; ALBREGT, H., and RATAN, R.: An effective method of placing air in the chest with CO_2 during open-heart surgery. J. thorac. cardiovasc. Surg. *53:* 618–622 (1967).
STRAUSS, R. H. and KUNKLE, T. H.: Isobaric bubble growth: a consequence of altering atmospheric gas. Science *186:* 443–445 (1974).
TANASAWA, I.; ECHIGO, R.; WOTTON, D. R.; NOMURA, M., and YANG, WEN-JEI: Measurements of mass diffusivity of gases in plasma and reaction velocity constant in bloods. Biomech. *4:* 265–273 (1971).
TAYLOR, G. I.: The formulation of emulsions in definable fields of flow. Proc. R. Soc. A *146:* 501 (1934).
TISOVEC, L. and HAMILTON, W. K.: Newer considerations in air embolism during operation. J. Am. med. Ass. *201:* 376–377 (1967).
TUCKER, R. W. and TENNEY, S. R.: Inert gas exchange in subcutaneous gas pockets of air-breathing animals: theory and measurement. Resp. Physiol. 1: 151–157 (1966).
US Navy Diving Manual (US Government Printing Office, Washington 1972).
VAN LIEW, H. D.: Coupling and diffusion and perfusion in gas exit from subcutaneous pocket in rats. Am. J. Physiol. *214:* 1176–1185 (1968).
VAN LIEW, H. D. and PASSKE, M.: Permeation of neon, nitrogen and sulfur hexafluoride through walls of subcutaneous gas pockets in rats. Aerospace *38:* 829–831 (1967).
YANG, WEN-JEI: Dynamics of gas bubbles in whole blood and plasma. J. Biomech *4:* 119–125 (1971).
YANG, WEN-JEI: A major cause of blood trauma in extracorporeal circulation; in Advances in bioengineering, pp. 167–168. (Am. Soc. Mechanical Engineers, New York 1974).
YANG, WEN-JEI, ECHIGO, R.; WOTTON, D. R., and HWANG, J. B.: Experimental studies of the dissolution of gas bubbles in whole blood and plasma. I. Stationary bubbles. II. Moving bubbles or liquids. J. Biomech. *4:* 275–281, 283–288 (1971).
YANG, WEN-JEI and LIANG, C. Y.: Dynamics of dissolution of gas bubbles or pockets in tissues. J. Biomech. *5:* 321–332 (1972).
YANG, WEN-JEI and MA, M. F.: Effects of dextran on dissolution of oxygen bubbles in whole blood. 1973 Biomechanics Symp., AMD – vol. 2, pp. 35–36 (Am. Soc. Mechanical Engineers, New York 1973).

Prof. WEN-JEI YANG, Mechanical Engineering Department, University of Michigan, *Ann Arbor, MI 48104* (USA)

Biomaterials and Interfacial Phenomena

J. FEIJEN, T. BEUGELING, A. BANTJES and C. TH. SMIT SIBINGA

Polymer Division, Department of Chemical Technology, Twente University of Technology, Enschede, and Coagulation Laboratory, Department of Internal Medicine, University of Groningen, Groningen

Contents

Abstract	100
I. Introduction	101
II. Characterization of Surface and Interface	101
III. Correlation of Interfacial and Surface Parameters of Materials with Phenomena Occurring after Contact of These Materials with Blood or Plasma	107
IV. Protein Adsorption onto Foreign Surfaces	108
A. Type of Protein Adsorption and Measuring Techniques	109
B. Protein-Interface Interactions	110
1. Apolar Surfaces	111
2. Polar Surfaces	111
V. Platelet Adhesion, Aggregation, and the Clotting of Blood	115
A. Platelet Adhesion and Aggregation	115
B. Blood Coagulation	116
C. Foreign Materials	119
VI. Concluding Remarks	126
Acknowledgements	126
References	127

Abstract

When a foreign material is exposed to blood, generally both platelet adhesion and activation of the intrinsic coagulation take place. The occurrence of these events depends on the properties of the contacting surface. Therefore, in section II of this chapter, some background information about concepts in surface chemistry, such as free surface energy, critical surface tension, wettability, work of adhesion and interfacial tension have been discussed. The relationship between surface parameters and the clotting of blood as proposed by several investigators has been critically reviewed in section III.

Several proteins from the blood adsorb onto a foreign material which is contacted with blood, hence the surface of such a material will change. A summary of the experimental methods used and a brief survey of some important variables and subjects in protein adsorption studies are presented in section IV. Some theoretical aspects concerning platelet adhesion, platelet aggregation and intrinsic coagulation are given in section V. Furthermore, experiments pertinent to both platelet adhesion and intrinsic coagulation on foreign materials are discussed. Conclusions and some suggestions related to the subjects mentioned above are stated in section VI.

I. Introduction

When a material is exposed to a biological environment, a contact between the material phase and the biological phase is established. The surface between these two phases is an interface. In this chapter we will deal with *the interface between foreign materials and blood*. Most of the events which will occur after contact of foreign materials with blood do originate at the interface. Therefore, it is very important to focus our attention on characterization of the interface and on interfacial phenomena [ANDRADE, 1973].

II. Characterization of Surface and Interface

The characterization of surfaces and interfaces requires some background information about concepts in surface chemistry [ADAM, 1968; ADAMSON, 1967], which will be provided here.

Free surface. When a solid or liquid is in contact with vacuum, the surface is called a 'free or ideal surface'. This surface can be characterized by a surface energy. *Surface energy* is a measure of the unsatisfied bonding capacity of the surface and can be due to unsatisfied primary bonds as well as to secondary bonds. Surface energy due to secondary bond interactions can be calculated by using dipole-dipole and London dispersion potention expressions [FOWKES, 1967; DE BOER, 1950]. For metals, the solid surface energy can be approximated from the heat of sublimation [BROPHY, 1964].

A molecule in the interior of a liquid is completely surrounded by other molecules and, therefore, on the average it is attracted equally in all directions. On a molecule at the surface, however, there is a resultant attraction inwards. When a surface area is enlarged, molecules are brought from the bulk into the surface, giving them a higher potential energy. At the same time the entropy of the system will change. This entropy change is not incorporated in the change of the surface energy. A better thermodynamic description of the system is, therefore, given by the surface free energy, equivalent to surface tension for free or ideal surfaces. The extended surface is in 'tension' and a surface can be described by its *surface tension*. When we are dealing with nondeformable solids, the occurrence of a surface tension can be precluded. The surface tension under specific external conditions is manifested by the tendency of contraction of the surface until the maximum possible number of molecules is in the interior. The unit of surface tension is $mN \cdot m^{-1}$ ($dyn \cdot cm^{-1}$), whereas the unit of surface free energy is $mJ \cdot m^{-2}$ ($erg \cdot cm^{-2}$).

Surface free energy or surface tension consists of surface energy and surface entropy. *For a liquid, the surface tension is defined as*

$$\gamma = \left(\frac{\partial G}{\partial A}\right)_{P,T} \quad \text{or} \quad \gamma = \left(\frac{\partial F}{\partial A}\right)_{T,V} \tag{1}$$

where G and F are the Gibbs and Helmholtz free energies. A is the surface area, and P, T and V are pressure, temperature and volume, respectively. The *surface tension of a liquid* can be measured as a force acting on a plate [DETTRE and JOHNSON, 1966] or by drop profile methods [WU, 1969; ROE, 1968; SAKAI, 1965; PATTERSON et al., 1971]. In the case of solids it is difficult to define the partial differential of equations 1, because the internal strain varies from point to point and it is necessary to know which variables have to be maintained constant.

The *surface tension of a solid* can be defined as the difference between the free energy of the bulk (F) and that of the surface (F'):

$$\gamma = \frac{F - F'}{area}.$$

Surface tension of solids cannot be measured directly. Some theoretical methods for estimating the surface tension of polymers from bulk properties are available. Using the free volume theory, one can obtain an expression relating surface properties to bulk properties [WU, 1974]. HILDEBRAND and SCOTT [1950] used an empirical relation between surface tension and the solubility parameter for small-molecule liquids. This equation was modified by WU [1968] and by PATTERSON and SIOW [1971] for polymers. ROE [1965] and SAFONOV and ENTELIS [1967] applied the parachor concept of SUDGEN [1924] for polymers. WU [1969, 1970, 1973] has used this relationship for many polymer systems. Furthermore, the corresponding state theory [PRIGOGINE et al., 1957] has been used by several workers [PATTERSON and SIOW, 1971; ROE, 1966; PATTERSON and RASTOGI, 1970] to correlate the surface tension of liquids and polymers, whereas STEWART and VON FRANKENBERG [1968] applied the significant structure theory.

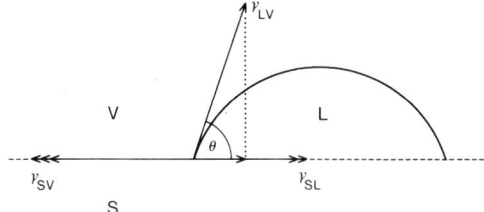

Fig. 1. Equilibrium spreading of a drop of liquid (L) on a solid surface (S) both in contact with vapor (V) of the liquid. The contact angle θ is determined by the three interfacial tensions γ_{SV} (solid-vapor), γ_{SL} (solid-liquid) and γ_{LV} (liquid-vapor).

Interface. When a phase terminates at the surface of another phase, an interface is obtained. The interfaces between solid-liquid and liquid-liquid phases are of major importance in the field of biomaterials. The *interfacial tension* (equivalent to interfacial free energy) is defined by

$$\gamma_{12} = \gamma_1 - \gamma_{1(2)} + \gamma_2 - \gamma_{2(1)} \qquad (2)$$

wherein γ_1 and γ_2 are the surface tensions of the phases 1 and 2, whereas $\gamma_{1(2)}$ and $\gamma_{2(1)}$ are expressing the effects of the phases on each other. The *work of adhesion* at an interface is a measure of the interface bonding and is defined as

$$W_{12} = (\gamma_1 + \gamma_2) - \gamma_{12} \qquad (3)$$

$$W_{12} = \gamma_{1(2)} + \gamma_{2(1)}. \qquad (4)$$

The work of cohesion W_c of a single phase 1 is $2\gamma_1$, because $\gamma_{11} = 0$ (no interface). When a drop of liquid is brought onto a solid surface, the equilibrium spreading of the liquid is given by the *contact angle equation* or *Young-Dupree equation* (fig. 1):

$$\gamma_{SV} = \gamma_{SL} + \gamma_{LV} \cos\theta \qquad (5)$$

wherein the interfacial tensions are given by γ_{SV} (solid-vapor), γ_{SL} (solid-liquid), and γ_{LV} (liquid-vapor). Here, γ_{SV} is defined as the surface tension of a solid phase in contact with vacuum, γ_S, minus the spreading pressure Π_S (a measure of the tendency of a vapor to adsorb and spread on the surface). γ_{LV} is defined in a similar way:

$$\gamma_{LV} = \gamma_L - \Pi_L.$$

The contact angle is given by θ. Equation 5 can be derived from a force-equilibrium for the interfacial tensions, or from a thermodynamic approach.

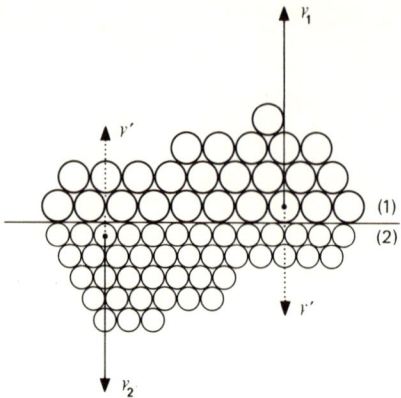

Fig. 2. Surface tensions acting between the immiscible phases 1 and 2; γ_1 and γ_2 are the surface tensions of phases 1 and 2, respectively, and γ' expresses the effect of the phases on each other.

In order to obtain more information about interfacial tensions we will discuss the *theory of* FOWKES [1964], and the *critical surface tension approach of* ZISMAN [1964].

For a liquid-liquid interface, where one liquid is not soluble in the other and only dispersion forces are acting between the phases, the interfacial tension is given by the following equation:

$$\gamma_{12} = \gamma_1 + \gamma_2 - 2\gamma' \qquad (6)$$

wherein γ_1 and γ_2 were defined before and γ' is the effect of one phase on the other (fig. 2). For dispersion force interactions, γ' will be a function of γ_1^d and γ_2^d. γ_1^d and γ_2^d are the dispersion contributions to the surface tensions of phase 1 and phase 2.

FOWKES postulated that

$$\gamma' = \sqrt{\gamma_1^d \gamma_2^d}. \qquad (7)$$

Combining equations 6 and 7 gives

$$\gamma_{12} = \gamma_1 + \gamma_2 - 2\sqrt{\gamma_1^d \gamma_2^d}. \qquad (8)$$

Measurements were carried out with alkane-water and alkane-mercury interfaces [WU, 1974]. For alkanes $\gamma \approx \gamma^d$. In these cases γ_1, γ_2 and γ_{12} can be determined. As a result, γ^d values for water and for mercury can be calculated. From these data, γ_{12} for water-mercury is calculated. The experimental value was in excellent agreement with the calculated value. Evidently, no polar interactions are acting between water and mercury.

In the case of *solid-liquid interfaces*, Fowkes' equation can be used to predict a contact angle. Combining equation 5 and equation 8 leads to

$$\cos\theta = -1 + 2\sqrt{\gamma_S^d} \cdot \frac{\sqrt{\gamma_L^d}}{\gamma_L}. \tag{9}$$

Again this equation only holds for the case where dispersion forces are acting between the two phases. Using equation 9, γ_S^d can be determined for solid surfaces (low energetic, apolar). For polar surfaces, this relation cannot be applied, because apolar liquids always spread on polar surfaces.

Sometimes a more general equation for the interfacial tension is used:

$$\gamma_{12} = \gamma_1 + \gamma_2 - 2\sqrt{\gamma_1^d \gamma_2^d} - 2\sqrt{\gamma_1^p \gamma_2^p} - 2\sqrt{\gamma_1^m \gamma_2^m} - \cdots. \tag{10}$$

The use of this equation can be criticized because the polar contributions γ_1^p, γ_2^p and the metallic forces γ_1^m, γ_2^m acting across the interface cannot be described by a geometric mean term. GIRIFALCO and GOOD [1957] proposed the following general relationship:

$$\gamma_{12} = \gamma_1 + \gamma_2 - 2\phi(\gamma_1 \gamma_2)^{0.5}. \tag{11}$$

ϕ can be estimated from molecular parameters for small-molecule liquids [GOOD and ELBING, 1970]. Such a calculation is not yet applicable to polymers. Combining equation 5 with equation 11 leads to the general form of *Good's equation* for the solid surface tension,

$$\gamma_S = \frac{[\gamma_L(1+\cos\theta) + \Pi_S]^2}{4\phi^2 \gamma_L} \tag{12}$$

wherein ϕ is a function of the molecular properties of the two phases (dipole moment, polarizability, ionization energy and molecular radius, etc). When the contacting phases are similar, ϕ will be 1. ZISMAN has correlated the *contact angle* of a homologous series of liquids on clean, low energy polymer surfaces with the *surface tension of the liquids* used (fig. 3). An empirical relation was found:

$$\cos\theta = a - b\gamma_L. \tag{13}$$

This linear relation between $\cos\theta$ and γ_L only holds for a limited range of surface tensions of liquids. Extrapolation to $\cos\theta = 1 (\theta = 0°)$ gives a *critical surface tension for wetting* γ_C. This critical surface tension is dependent on the surface and the homologous series of liquids used. For low energy surfaces, γ_C values for different homologous series of liquids are equal.

The *critical surface tension of a solid* has been correlated with the *specific surface free energy* [ZISMAN, 1964]. The correlation between γ_C and γ_S is often very good [OWENS and WENDT, 1969]. It is, however, *dangerous to correlate critical surface tensions with phenomena occurring at*

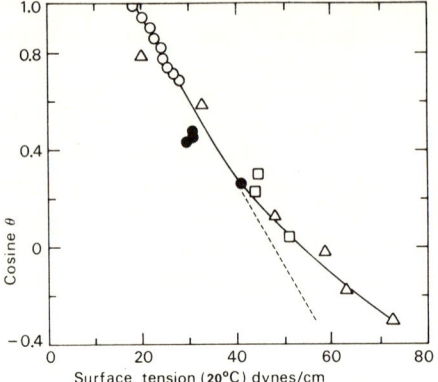

Fig. 3. Wettability (cosine θ) of polytetrafluoroethylene (Teflon) as a function of surface tension of various liquids; ○ = n-alkanes; ● = esters; □ = nonfluoro-halocarbons; △ = miscellaneous liquids. The critical surface tension γ_C is obtained by extrapolating to cosine θ = 1. It can be seen that γ_C depends on the series of homologous liquids used. [Redrawn from FOWKES, 1964.]

the interface. The values of γ_C for different solid surfaces depend on the series of homologous liquids used and, therefore, γ_C is not a good parameter for the characterization of a surface.

In conclusion, interfacial tensions for liquid-solid systems cannot be measured. For liquid-liquid systems the interfacial tension can be determined [WU, 1969; ROE, 1968; SAKAI, 1965]. Estimations for the dispersion and polar components of the surface free energy for a variety of polymer surfaces have been made by KAELBLE [1970]. Using these data and equation 10, values for interfacial tensions of polymer-water interfaces have been calculated by ANDRADE [1973]. The data given by KAELBLE, however, can be criticized. KAELBLE used the following relations for the description of the interactions of two nonwetting liquids to a common solid surface:

$$(W_A/2)_i = (\gamma_L^d)_i^{1/2}(\gamma_S^d)^{1/2} + (\gamma_L^p)_i^{1/2}(\gamma_S^p)^{1/2}$$
$$(W_A/2)_j = (\gamma_L^d)_j^{1/2}(\gamma_S^d)^{1/2} + (\gamma_L^p)_j^{1/2}(\gamma_S^p)^{1/2}. \tag{14}$$

These equations define the case of characterized liquids (i) and (j), interacting with a common solid (S) so as to provide two sets of values for $W_A/2$ (W_A is the work of adhesion between a liquid and a solid surface), γ_L^d and γ_L^p; γ_S^d and γ_S^p may then be calculated. However, the value of γ_S^p will depend on the liquid used for the measurements. Furthermore, it is doubtful whether the polar interactions can be described by a geometric mean term. LYMAN *et al.* [1965] calculated the surface free energy of polymers from contact angle measurements using Good's equation 12. These authors assumed ϕ in this formula to be 1. This assumption is only allowed when the polymer surface and the contacting liquid do have similar properties. In the case of an apolar surface and water ϕ will not be 1 and, therefore, some of their values are not correct. Realizing this, ANDRADE et al. [1973] generated a set of data for the interfacial tensions, using the data of LYMAN et al. and using equation 11 (ϕ = 1). At this moment *only rough estimations* of interfacial tensions for a solid-liquid system can be made.

Effects of charge and surface potential have not been included in the foregoing considerations. The correlation of charge and surface potential with blood clotting is discussed, in this book, in the chapter by SRINIVASAN et al.

Polymer surfaces. The surface characteristics of a polymer are determined by the types of chemical bonds and atoms present on the surface, the density of the surface [ROE, 1965], and the crystallinity and orientation of the surface [SCHONHORN, 1965, 1967, 1968; SCHONHORN and RYAN, 1966, 1969]. Surface properties can be modified by adsorption of gases and liquids. Also mold-release agents, plasticizers and other additives can modify the surface properties. Finally, surface reactions can cause a different chemical composition of the material at the surface than in the bulk. In order to compare the surface properties of a series of materials it is necessary to specify the precise chemical composition, synthetic pathway, processing methods and other treatments.

III. Correlation of Interfacial and Surface Parameters of Materials with Phenomena Occurring after Contact of These Materials with Blood or Plasma

Water wettability of solid surfaces has been related to clotting times of fresh animal blood. Lampert's Rule (clotting time is inversely proportional to water wettability) has been proposed in connection with the differences in clotting times of blood in glass tubes and in siliconized glass tubes. Later, the surface free energy of a solid surface was chosen by LYMAN et al. [1965] as a criterion for blood clotting properties and for platelet adhesion [LYMAN et al., 1968, 1970a].

The work of adhesion at the polymer-blood interface was related to the clotting time of blood by BISCHOFF [1968]. Biocompatibility of materials was correlated with the critical surface tension of polymers by BAIER [1972]. Polymers having a value of γ_C of about 25 dyn·cm^{-1} were considered to be the most biocompatible materials.

ANDRADE [1970] suggested that the interfacial free energy is the most important property. This suggestion has been the subject of several papers [ANDRADE et al. 1971, 1972, 1973]. At this moment, from the concepts mentioned above *only the minimal interfacial free energy concept seems to be attractive for the correlation with biocompatibility*. Materials which can minimize the interfacial free energy in contact with water are, for instance, *hydrogel type materials* [ANDRADE, 1973; WICHTERLE and LIM, 1960].

A variety of hydrogel type materials have been synthesized, and in general these materials do have a very good compatibility with blood and with tissue. Hydrogels are cross-linked three-dimensional networks. These materials do swell in water, and different equilibrium amounts of water can be incorporated depending on the type of material and the amount and type of cross-links. Using ANDRADE's concept, the ideal material will contain large amounts of water. Detailed studies have to be carried out to correlate the amounts of water in the gel (interfacial free energy) with compatibility, especially in the regions where the gel contains more than 90% water.

Increasing amounts of water in these gels lead to one serious drawback of this approach, namely, a decrease in mechanical properties. *Applicable biomaterials* can however, be obtained by *hydrogel surface coating methods* [LEE et al., 1972].

The interfacial tensions estimated for different polymer-water systems have been plotted by ANDRADE as a function of the free surface energy of the polymer surfaces. The curves show a minimum of $\gamma_{12} = 0$ dyn·cm^{-1} for the surface free energy of water ($\gamma = 73$ erg·cm^{-2}). Glass and nylon do have a γ_{12} value in contact with water of 24 and 20 dyn·cm^{-1}, respectively; polydimethylsiloxane (PDMS) has a γ_{12} of 35 dyn·cm^{-1}. It is well known that PDMS has a better blood-compatibility than glass or nylon.

If the values for the interfacial energies are correct, we can conclude that *the minimal free interfacial energy concept is more important in a small range of low free interfacial energies* (for instance, hydrogels containing a large amount of water); furthermore, that the free interfacial energy is only one parameter in the set of surface parameters, which will determine the ultimate behavior of a material in a biological environment.

IV. Protein Adsorption onto Foreign Surfaces

When a material is brought in contact with blood, the first event which takes place is the adsorption of proteins from the blood onto the material surface. An extensive amount of work has been carried out by several authors to obtain more insight into these phenomena [FALB et al., 1967; BRASH and LYMAN, 1971; VROMAN et al., 1971; SALZMAN et al., 1969; BAIER et al., 1971; LEE and KIM, 1974a]. Upon adsorption of plasma proteins, the surface of the material will be altered and the subsequent events will be determined by the modified surface. In order to obtain a better understanding of the behavior of materials in contact with blood it is necessary to investigate protein adsorption phenomena. Many protein adsorption experiments have been carried out under no-flow conditions. However, some work is available where controlled flow conditions have been applied [LEE and KIM 1974a; LYMAN and KIM, 1971]. *Protein adsorption phenomena are important for both the triggering of the intrinsic coagulation process and for platelet adhesion.* These two important subjects will be described in section V. This section contains some background information about material-protein adsorption phenomena and a brief survey of experimental techniques used to study these phenomena.

A. Type of Protein Adsorption and Measuring Techniques

Protein adsorption at a solid-liquid interface can frequently be described by the *Freundlich equation* [KIPLING, 1965]:

$$\frac{x}{m} = Kc^h \tag{15}$$

or the *Langmuir equation* [ZISMAN, 1964]:

$$\frac{x}{m} = \frac{ac}{1+bc} \tag{16}$$

wherein the quantity of solute adsorbed at the equilibrium concentration c is x per amount of solid weight m; K, h, a and b are experimental constants. Both equations 15 and 16 can only be used at low concentrations of the solute. The Langmuir isotherm is the commonest type of adsorption isotherm observed, and this isotherm usually holds for a somewhat more extended concentration range than does the Freundlich isotherm.

Most of the experimental work on protein adsorption has been carried out with albumin, γ-globulin and fibrinogen, using different types of surfaces. The techniques used for these investigations are solution depletion techniques and direct study of the surface. One of the direct techniques used is infrared (IR) reflection spectroscopy [BRASH and LYMAN, 1969, 1971]. The amide A, amide I and amide II bands are chosen for the detection of the protein concentration at the surface. The thickness of protein layers and the average area per molecule can be estimated from the surface concentration if monolayers are assumed. On the basis of the calculated dimensions, BRASH and LYMAN [1971] concluded that plasma proteins are not dimensionally denatured by adsorption to some hydrophobic surfaces and that the protein layer was a closed-packed layer adsorbed on end. Using this technique, the surfaces have to be dried in order to measure the IR spectra; this may alter the conformation of the adsorbed layer. Furthermore, it is not possible to distinguish different proteins on the surface, adsorbed from a protein mixture solution.

Another approach was applied by LEININGER and co-workers using zeta-potential [LEININGER et al., 1966] and radiolabeling of proteins [FALB et al., 1967]. The zeta-potential change of polymer surfaces was studied after protein adsorption. This technique cannot be applied to determine surface concentrations of proteins. Radiolabeling of proteins has also extensively been applied by KIM and co-workers [LEE and KIM, 1974a,b; LEE et al., 1974] and by other authors [LEMOS et al., 1974]. This method is also suitable to study *adsorption of mixtures of proteins onto solid surfaces*. VROMAN et al. [1969a,b] carried out extensive studies using *ellipsometry*. The technique gives values for the thickness and surface concentration of protein layers at the surface.

Electron microscopy was also used as a technique to study adsorbed proteins on different surfaces [HALL, 1956; VALENTINE, 1959]. DILLMAN and MILLER [1973] applied colorimetric methods, whereas MCMILLIN and WALTON [1974] applied a circular dichroism technique. This technique was used for the detection of *conformational changes of adsorbed proteins*. Recently, MORRISSEY and STROMBERG [1974] studied the conformation of adsorbed blood proteins by infrared bound fraction measurements.

B. Protein-Interface Interactions

The interaction of proteins with different surfaces is governed by both kinetic and thermodynamic factors. HOFFMAN [1974] developed a general approach distinguishing two cases: (a) hydrophobic, apolar, non-wettable surfaces and (b) hydrophilic, polar, wettable surfaces. When protein adsorption takes place, the change in free energy will be negative. This means that the process is governed by the changes in enthalpy and entropy.

1. Apolar Surfaces

Water molecules at the surface will be highly structured [DROST-HANSEN, 1971]. If charges are available on the surface, the interface will have water regions similar to hydration shells of ions. A diffuse double layer of counterions will be set up. Upon adsorption of a protein, the interacting forces will be mainly nonpolar or dispersion forces. Ice-like water, which is structured in the vicinity of the interface and the apolar parts of the protein molecule, will be released, effecting a gain of entropy. Some of the charges at the surface will be neutralized by charges on the protein molecule, releasing bound counterions from both the interface and the protein molecule (gain in entropy). Some degree of unfolding will take place when protein molecules expose their apolar moieties to the apolar surface. This means that the conformation of a protein molecule will change (more or less denaturation).

2. Polar Surfaces

Water molecules at polar surfaces will be oriented to some extent by dipole-dipole interactions, and if ions are available at the surface, some of the water in the interface will have a similar structure as the water in hydrated ions. Adsorption of protein molecules will take place by ion-ion or/and dipole-dipole interactions. Again oriented water molecules will be released from both the surface and the protein molecule. In this case, the change of the conformation of the protein molecules will in general be less as compared to the adsorption process onto apolar surfaces. Generally, protein molecules can be desorbed better from polar surfaces as compared with apolar surfaces. We will now discuss some results of work on protein adsorption in the light of the foregoing considerations.

At this moment many investigations by several authors, using the different techniques already mentioned, have been carried out. However, several important questions remain unanswered. Many protein adsorption studies have been performed in the presence of an air-water interface. This might change the conformation of proteins adsorbed at the interface. When this protein layer contacts the surface which is studied, erroneous results can be obtained [LEE and HAIRSTON, 1971].

Human as well as bovine proteins have been applied, and it is very well possible that different results will be obtained when the same surfaces and similar experimental conditions are used.

Protein adsorption experiments, using one protein in solution, do not yield valuable information for the case where a mixture of proteins (plasma or blood) *can compete for adsorption onto the surface.* This is illustrated by the work of VROMAN et al. [1971]. These authors studied the effect of protein solutions and plasma (intact or activated) on already adsorbed films of human albumin, 7S γ-globulins and fibrinogen onto oxidized silicon crystal slices, using ellipsometry. Albumin films (15–25 Å) gained thickness after exposure to all proteins and plasmas. They lost some antigenicity to anti-albumin and gained the antigencity of the protein to which they had been exposed. Globulin films were 50–65 Å thick, lost thickness on exposure to albumin, gained on exposure to fibrinogen and intact plasma. Globulin films did not loose antigenicity, but after exposure to fibrinogen they gained activity to react with anti-fibrinogen serum. Fibrinogen films (70–80 Å) lost thickness on exposure to all proteins, sera, and plasma except activated plasma, which caused an increase. Antigenicity of fibrinogen films was destroyed by intact plasma only and markedly reduced by activated plasma.

LEE et al. [1974] have studied competitive adsorption of plasma proteins onto different hydrophobic neutral surfaces, using radiolabeled proteins (bovine albumin, γ-globulin and fibrinogen). The polymer substrates were silastic rubber (SR), fluorinated ethylene-propylene copolymer (FEP) and a segmented copolyether-urethane-urea based on polypropylene glycol (mol. wt. 1,025), methylene bis (4-phenylisocyanate) and 1,2-diamino ethane (PEU). Air-water interfaces were avoided. After plateau values were reached, the protein layer on the different surfaces consisted of 43–45% fibrinogen. The values for γ-globulin were: FEP, 18%; SR, 24%, and PEU, 17%, and for albumin: FEP, 39%; SR, 32% and PEU, 38%. The plateau times for γ-globulin and fibrinogen in the case of FEP were reached within 1 min. The plateau times in the case of adsorption of a single protein (fibrinogen or γ-globulin) on FEP were considerably longer (15 and 20 min). No straight-forward explanation for this behavior has been given. At this time, these hydrophobic surfaces have not been treated first with a solution of a single protein, followed by exposing the proteinated surface to other protein solutions, plasma and blood, as was carried out by VROMAN et al. [1971] for a hydrophilic surface. Then it should be possible to compare the behavior of single proteins on hydrophobic as well as on hydrophilic surfaces by comparison of their subsequent behavior upon exposure to other protein solutions.

BRASH and LYMAN [1971] compared protein adsorption onto hydrophobic surfaces with the adsorption onto hydrophilic surfaces. For hydrophobic surfaces, adsorption is generally irreversible. Albumin and γ-globulin are completely desorbed from a hydrophilic surface like cuprophane by immersion in water. These differences suggest that *proteins interact less with neutral hydrophilic than with hydrophobic materials.*

Some evidence has been forwarded that smaller amounts of protein adsorb onto hydrophilic materials, as compared with hydrophobic materials [ANDRADE, 1973; BRASH and LYMAN, 1971]. A dependence of the amount of adsorption on the molecular weight of proteins was found in the case of adsorption onto hydrophilic surfaces [BRASH and LYMAN, 1971]. This relationship was not found with hydrophobic materials.

At this moment, there is not much information about the change in conformation of proteins after adsorption at hydrophobic or hydrophilic surfaces. MORRISSEY and STROMBERG [1974] determined the amount of bound fraction of albumin, fibrinogen and prothrombin (bovine) onto a silica surface, using IR spectroscopy. It was concluded that for albumin and prothrombin the internal bonding of these globular proteins appears sufficient to prevent any gross unfolding of the molecules and that the native structure was maintained, while adsorbed. Fibrinogen shows some interfacial aggregation at the surface with increasing adsorbance. BRASH and LYMAN [1971] conclude that *plasma proteins are not dimensionally denaturated by adsorption onto hydrophobic surfaces.*

The change of conformation of bovine fibrinogen and human clotting factor XII (Hageman factor) after adsorption onto a quartz substrate was studied by MCMILLIN and WALTON [1974] using circular dichroism (CD) techniques. Fibrinogen has the same or a similar conformation as compared to the conformation in aqueous solution, but two-dimensional ordering causes very minor CD spectroscopic changes. Factor XII shows a large change of conformation upon adsorption. A possible role of water in the activation process of factor XII was proposed. Much more data are needed to obtain insight in conformational changes of proteins, adsorbed at polymer surfaces.

LYKLEMA and NORDE [1973] studied the adsorption of human albumin onto polystyrene (latex systems). It was concluded that *adsorption as a function of pH was maximal in the isoelectric point* (IEP); in this case adsorption takes place side-on with native human serum albumin molecules. Below and above the IEP, adsorption proceeds in a more unfolded conformation but no spread monolayer is observed. Several investigators also found that adsorption of human serum albumin and comparable proteins is maximal in or close to the IEP for different hydrophilic surfaces, pyrex glass [BULL, 1965] and glass [MCRITCHIE, 1972]. It has to be stressed that protein adsorption is dependent on buffers, electrolytes, pH and surface charge.

DILLMAN and MILLER [1973] have carried out adsorption experiments using bovine serum albumin, γ-globulin and fibrinogen on the materials cellophane, silicone rubber, silicone rubber-Lexan copolymers and on a series of ion exchange membranes. The adsorption was followed by colorimetric methods. These authors come to a very interesting conclusion, namely, adsorption of proteins onto polymer membranes takes place via two separate and distinct ways. Both types are Langmuir types of adsorption and take place simultaneously without interaction. The first type of adsorption is characterized as being relatively hydrophilic, easily reversible, with a heat of adsorption characteristic of a condensation process. The second type is characterized as being highly bound, hydrophobic and with an endothermic heat of adsorption. Platelet adhesion onto foreign surfaces has been related to protein adsorption by PACKHAM et al. [1969], ZUCKER and VROMAN [1969], SALZMAN et al. [1969] and KIM et al. [1974]. It turns out that surfaces coated preferentially with γ-globulin or fibrinogen do adhere more platelets as compared with surfaces coated with albumin. A mechanism has been proposed by KIM et al. [1974] and LEE and KIM [1974b]. This will be discussed in section V.

Most authors suppose that wettable, negatively charged foreign surfaces can initiate the intrinsic coagulation by adsorption and activation of factor XII (see section V). Activation of factor XII (a sialoglycoprotein) on glass can be inhibited by protamine sulfate. Two possible explanations come to mind. Protamine sulfate will render glass more positive, preventing the adsorption of factor XII. Furthermore, protamine sulfate may prevent a conformational change of factor XII, necessary for its activation [VROMAN, 1967]. VROMAN [1967] has carried out experiments indicating that factors XI and XII are involved in the desorption of fibrinogen from a protein layer adsorbed at TaW and SiW slides in contact with plasma. Various kinds of glass have also been exposed to plasma and blood [VROMAN et al., 1971]. After various exposing times, the surfaces were rinsed with buffered saline. The surface coatings were investigated with antisera and it was shown that blood or plasma, deposits a film of fibrinogen (50 Å thick) within a few seconds. Then, if the plasma is intact (not activated), it converts, masks, or replaces this film within 20 sec. If the plasma is not activated, part of the film is removed. Until now no answer is available as to where factor XII is activated among the fibrinogen molecules. Experiments carried out with fibrinogen-deficient plasma do not show a relationship between the presence of fibrinogen and the activation of factor XII.

In conclusion, *one of the best approaches in biomaterials may be the minimal interfacial energy approach, leading to minimal or no protein adsorption and possibly no activation of factor XII.* Surfaces preferentially coated with albumin can be successful in respect to minimal platelet adhesion. However, the activation of factor XII is not clearly understood, preventing straightforward strategies to combine the inhibition of platelet adhesion and activation of factor XII.

V. Platelet Adhesion, Aggregation, and the Clotting of Blood

One of the major problems in the application of biomaterials is the interaction of such materials with blood. This interaction may lead to a number of events [BRUCK, 1973a] such as thrombus formation and the destruction of erythrocytes (hemolysis), leukocytes and platelets. Up to now, the long-term prevention of thrombus formation on the surface of a foreign material has been most difficult to achieve, in the development of biocompatible materials.

In normal hemostasis as well as in thrombosis a number of reactions occur in which platelets and coagulation factors, present in the blood, are involved. Much of this knowledge has been gathered from *in vitro* experiments in which foreign materials have been used. Even if such experiments might bear little relationship to what really happens *in vivo*, they become important when foreign materials are introduced into the blood stream [CHESSELLS and BLACK, 1972].

The clotting of blood in a test tube can perhaps be compared to some extent with the formation of a venous thrombus; cellular elements from the blood are more or less distributed at random throughout a fibrin clot which is formed [WEISS, 1974]. By contrast, the adhesion of platelets to a damaged vascular wall and their subsequent aggregation play a major role in the formation of an arterial thrombus, and fibrin is formed in a later stage [WEISS, 1974]. It is generally accepted that fibrin is formed at the end of a cascade process in which coagulation factors are involved.

There is evidence that *in normal hemostasis, platelets participate in reactions with coagulation factors at every stage of the blood coagulation process* [WALSH, 1974]. For the sake of simplicity, we shall distinguish in the following discussion between the role of the platelets and coagulation factors in the clotting of blood. We will not discuss a possible role of the leukocytes in hemostasis [GUEST *et al.*, 1973]. More detailed information about the function of platelets and coagulation factors in the clotting of blood may be found in WILLIAMS *et al.* [1972], HEMKER *et al.* [1969], and WALSH [1974].

A. Platelet Adhesion and Aggregation

The adhesion of platelets to an injured vessel wall or to foreign surfaces also has been called adhesiveness, adherence or agglutination by different authors. We shall call the property of the platelets to adhere to an injured vessel wall or to a foreign surface 'adhesion'. 'Aggregation' is used as a general term for the attachment of platelets to each other.

It is well known that *platelets rapidly adhere to the connective tissue of a damaged vascular wall* [ZUCKER, 1972]. *In vitro*, they also adhere and aggregate in the presence of connective tissue suspensions or collagen suspensions [ZUCKER, 1972; HOVIG et al., 1968; SPAET and LEJNIEKS, 1969]. Therefore, *the primary step in hemostasis is thought to be the adhesion of platelets to exposed collagen fibers.* This adhesion is probably due to the formation of an enzyme-acceptor complex between the incomplete carbohydrate chains (galactosyl residues) of collagen and an enzyme of the platelet membrane [glycosyl transferase; JAMIESON, 1973].

Platelets which are in contact with collagen fibers contract and release adenosine diphosphate (ADP) and numerous other platelet constituents. ADP causes more platelets to adhere to those already adhering. In this way an autocatalytic process is started in which platelet aggregates are formed. The coagulation process (see below), which also takes place, results in the formation of thrombin. Thrombin, being a release inducer, also causes platelet aggregation. Ultimately, platelets irreversibly sticked together to large aggregates do form the so-called hemostatic plug.

Another mechanism must, however, account for the adhesion of platelets to foreign surfaces, because no collagen is present in these cases. We will deal with this subject in section V.C.

B. Blood Coagulation

A number of reactions, mainly occurring between soluble proteins in blood or plasma, which lead to the formation of insoluble fibrin, are called *blood or plasma coagulation*. The substances which are involved in these reactions are called *coagulation factors*. Table I summarizes the accepted Roman numerals, together with synonyms, for these factors [WILLIAMS, 1972]. Except factor III (tissue factor), the numerals denote the factors as they are present in blood or plasma. Factors which have been converted to enzymatically active forms (except fibrin, which is not enzymatically active) are indicated with a Roman numeral to which the suffix 'a' is added. Generally, factors I and IIa are called fibrinogen and thrombin, respectively. An important substance in blood coagulation is a phospholipid which is released by the platelets; this factor is not represented by a Roman numeral and is known as platelet factor 3.

MACFARLANE [1964] and DAVIE and RATNOFF [1964] have proposed a coagulation scheme, known as the cascade system or waterfall sequence, in which is stated that coagulation factors are sequentially activated. Later experiments indicated that the original scheme was not correct, and some

Table I. International nomenclature for blood coagulation factors

Factor	Synonym(s)
I	fibrinogen
II	prothrombin
III	tissue factor, tissue thromboplastin
IV	Ca^{2+}
V	proaccelerin
(VI)	not assigned, sometimes called: accelerin, identical to activated factor V (Va)
VII	pro-convertin
VIII	antihemophilic factor A (AHF-A)
IX	antihemophilic factor B (AHF-B), Christmas factor, plasma thromboplastin component (PTC)
X	Stuart-Prower factor
XI	plasma thromboplastin antecedent (PTA), antihemophilic factor C
XII	Hageman factor
XIII	fibrin-stabilizing factor, fibrinase, Laki-Lorand factor

changes were introduced. The coagulation scheme given in figure 4 is based on the scheme proposed by HEMKER and KAHN [1967].

In vitro, the so-called intrinsic pathway (the coagulation occurring as a result of the interaction of factors present in the circulating blood) starts with the activation of factor XII by exposure to a foreign surface. This activation is brought about by the adsorption of factor XII onto the surface. Factor XIIa (activated factor XII) functions as a proteolytic enzyme in activating factor XI. The reaction between factor XIIa and factor XI seems to take place on the surface [HAANEN, 1969].

In vivo, *platelets most probably play an important role in the triggering of the coagulation process* [WALSH, 1974]. There is evidence that two alternative pathways exist [WALSH, 1972a,b, 1974]. Firstly, when stimulated by ADP, platelets can activate factor XII [WALSH, 1972a]. Secondly, platelets adhering to collagen can initiate the coagulation by activating molecules of factor XI, which are platelet-associated [WALSH, 1972b]. In normal hemostasis, the second pathway is probably more important than the activation of factor XII [WALSH, 1974], because individuals with Hageman trait function hemostatically well. Thus, there seems to be an important difference between the triggering mechanisms of the *in vivo* and *in vitro* coagulation processes.

Factor XIa appears to function as an enzyme in the activation of factor IX; Ca^{2+} is needed for this reaction. The formed factor IXa

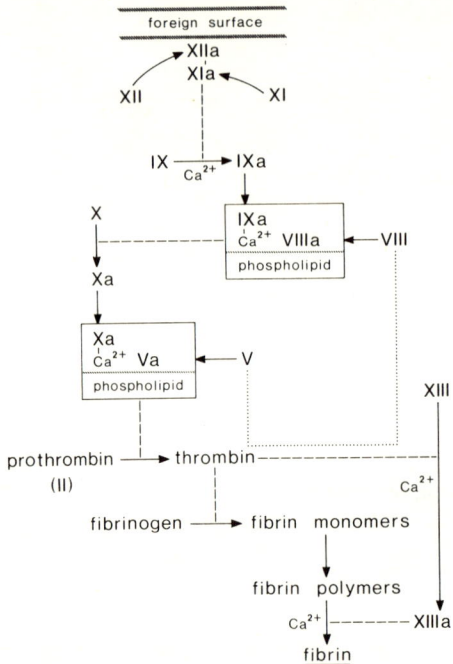

Fig. 4. Activation of coagulation factors after exposure of a foreign surface to blood or plasma. The conversion of a factor or a reaction product is indicated with a solid arrow (e.g., factor XIII→factor XIIIa). The broken lines indicate catalytic functions (e.g., thrombin catalyzes the conversion of fibrinogen→fibrin monomers and factor XIII→factor XIIIa). The dotted lines indicate the action of thrombin on factors VIII and V to convert them to a more reactive form. Complexes of different factors are given by the two rectangles. The phospholipid of these complexes may be provided by platelets (platelet factor 3 or membrane lipid). In normal hemostasis, platelets probably provide a catalytic surface for all the reactants given in the scheme. Platelets adhering to collagen can initiate the coagulation by activating platelet-associated factor XI (bypassing factor XII) and/or they can activate factor XII in the presence of ADP [see text and WALSH, 1974].

participates in the formation of a complex in which Ca^{2+}, factor VIIIa, and a phospholipid are involved. This so-called complex II converts factor X to factor Xa. Factor VIIIa may be formed from factor VIII by traces of thrombin. The phospholipid may be provided by the platelets (platelet factor 3) during the release reaction. There is also evidence that the intact platelet surface catalyzes the reactions of factors XIa, VIII, IX and X in the formation of factor Xa [WALSH and BIGGS, 1974].

In a similar way, as was pointed out for factor IXa, factor Xa together with Ca^{2+}, factor V (or Va), and phospholipid form a complex (prothrombinase, or complex I), which converts factor II (prothrombin) into the enzyme thrombin. Once thrombin has been formed, a number of reactions occur, ultimately converting the soluble fibrinogen from the plasma into an insoluble fibrin network.

We will not discuss the so-called *extrinsic system*, in which the coagulation process starts by the action of a tissue factor (factor III).

C. Foreign Materials

In section V.B. it has been mentioned that the *in vitro* coagulation process is triggered by the activation of factor XII on a foreign surface. It is well known that this activation occurs at plasma-solid interfaces, in which intact platelets are not involved. The contact activation of factor XII depends on the physicochemical properties of the solid material used. Furthermore, we mentioned that, in normal hemostasis, a mutual interaction between the platelets and coagulation factors may be essential for the coagulation process. In this case, the chemical nature of the exposed collagen surface probably plays an important role. We will now continue with the processes which occur on the surface of a foreign material coming into contact with living blood.

Generally, *exposure of blood to a foreign material causes both platelet adhesion and coagulation*. Already in 1927, SHIONOYA and ROWNTREE made an extravascular loop of collodion between the arteria carotis and vena jugularis (A-V shunt) of a rabbit and observed the formation of platelet thrombi, followed by coagulation. Only white thrombi were formed when the animals were injected with heparin. Consequently, platelet adhesion is not inhibited by heparin dissolved in the blood. This conclusion is confirmed by experiments of LEONARD and FRIEDMAN [1970] and of REMBAUM et al. [1973]. REMBAUM et al. inserted chronic A-V shunts, prepared from silicone rubber and polyurethane, in beagles. Massive doses of heparin (4 mg heparin/kg body weight) did not affect platelet adhesion onto the polymer surfaces. Heparin inhibits the action of factors XIa, Xa, and thrombin [DAMUS et al., 1973]. Therefore, platelet adhesion onto polymer surfaces is possible before a platelet-aggregating substance such as thrombin has been formed via the intrinsic pathway. This can also be concluded from the findings of MASON et al. [1972]. They found that native blood from patients congenitally deficient in factors VIII or IX showed *no reduced* platelet adhesion onto glass, siliconized glass and teflon as compared to native blood from a normal subject.

It was already mentioned in section IV that platelet adhesion onto foreign surfaces has been related to protein adsorption by several authors [SALZMAN et al., 1969; PACKHAM et al., 1969; ZUCKER and VROMAN, 1969; KIM et al., 1974]. *Foreign materials, when exposed to plasma or blood, rapidly adsorb plasma proteins* [FALB et al., 1967; BRASH and LYMAN, 1971; VROMAN et al., 1971; SALZMAN et al., 1969; BAIER et al., 1971; LEE and KIM, 1974a,b]. Up to now, the composition of the protein layer(s), adsorbed from blood onto foreign materials, is largely unknown. VROMAN and ADAMS [1969a], using an ellipsometer, found that oxidized silicon crystals adsorbed fibrinogen out of plasma within 2 sec. The adsorbed protein was identified by its ability to bind fibrinogen antiserum. Glass, mainly consisting of SiO_2, promotes the adhesion of platelets, and therefore it was suggested [ZUCKER and VROMAN, 1969] that fibrinogen, while it was bound to glass, might cause platelet adhesion.

In a further study, ZUCKER and VROMAN [1969] showed that platelets adhered to glass slides previously exposed to normal plasma or to a fibrinogen solution, but not to slides exposed to serum or afibrinogenemic plasma. VROMAN and ADAMS [1969a] found no detectable amounts of fibrinogen on heparin-treated GMAC (in which heparin is linked ionically to GMAC, a polymer of 2-hydroxy-3-methacryloyl-oxypropyl-trimethyl ammonium chloride) which has been contacted with plasma. After contact of the same heparinized surface with platelet-rich plasma or blood no platelet adhesion could be observed. STONER et al. [1971] found that mica on which heparin had been adsorbed did not adsorb fibrinogen from fibrinogen solutions or plasma. The mechanism of this inhibitory effect is most probably similar to that of the inhibition of fibrinogen adsorption on the heparin-GMAC complex found by VROMAN and ADAMS [1969a]. STONER et al. explained the inhibition of the fibrinogen adsorption on heparinized mica in terms of a charge phenomenon.

Heparinized surfaces also diminish platelet adhesion *in vivo*. For instance, in the A-V shunts used by REMBAUM et al. [1973], platelet adhesion onto a polyurethane-heparin complex was less than onto a commerical polyurethane.

PACKHAM et al. [1969] investigated the effect of glass surfaces, coated with plasma proteins, on ^{14}C-serotonin-labeled platelets in platelet-rich plasma. The greatest amount of radioactivity became associated with glass surfaces which had been exposed previously to a fibrinogen or γ-globulin solution. Less radioactivity was found on albumin-coated glass (even less than glass). Obviously, fibrinogen and γ-globulin coating promotes platelet adhesion to the surface.

Later experiments of VROMAN [1971] showed that ultrathin cellulose membranes adsorbed fibrinogen out of plasma at air-plasma-membrane rather than at plasma-membrane interfaces, and this adsorbed fibrinogen attracted platelets. In initial studies of blood-polymer interactions, LYMAN et al. [1968, 1970b] found that *heavy platelet adhesion occurred when a polymer surface was moved through a blood-air interface.* Therefore, it was suspected that the polymer was coated with a denatured protein layer and that this layer caused platelet adhesion. In a further investigation, LYMAN et al. [1970] examined the effect of 'native' protein layers on platelet adhesion. Several polymers were coated with albumin, fibrinogen, γ-globulin and plasma, respectively. Air-solution or air-plasma interfaces were avoided in the coating procedures. Venous blood, drawn directly from a human donor, had no contact with air when it flowed along a coated polymer. Their results show that fewer platelets adhered to a polymer precoated with a layer of one of the proteins (or plasma) than to the pure polymer surface, or a surface which had been exposed to a blood-air interface. Fibrinogen- and γ-globulin-coated surfaces adhered somewhat more platelets than those coated with albumin. Increasing exposure times (ranging from 30 sec to 3 min) did not increase the adhesion of platelets on a fluorinated ethylene-propylene copolymer precoated with native albumin. The uncoated copolymer showed an increase with a factor of about 7. This experiment indicates that blood coates the untreated polymer with one or more proteins, which promote platelet adhesion.

In a similar investigation as was carried out by LYMAN et al., MASON et al. [1973] studied the effect of human blood on glass surfaces precoated with various purified proteins. They used an *ex vivo* test cell in which the protein solution was replaced by saline. Then saline was displaced by native blood in the absence of a blood-air interface, and the blood remained undisturbed in the cell for 5 min. Their data show that only glass which was precoated with fibrinogen markedly promoted platelet adhesion. In the experiments of MASON et al. [1972] with native blood from patients congenitally deficient in a coagulation factor, mentioned above, the same type of an *ex vivo* test cell was used, and blood-air interfaces were also avoided. During these investigations they found that *deficiency of fibrinogen markedly reduced adhesion of platelets to test surfaces of glass, siliconized glass and teflon.*

Recently, LEE and KIM [1974b] used *Silastic rubber tubes*, coated with different plasma proteins, as flow-through cells for blood taken directly from the jugular viens of anesthetized, unheparinized sheep. Air-blood interfaces were avoided by filling the tubes with buffered saline. After 3 min of flow, they found *more platelets adhering onto the fibrinogen-coated tubes than onto the γ-globulin-coated ones; albumin-coated tubes showed less platelet adhesion.*

To get more insight in the nature of a protein layer adsorbed from blood, LEE et al. [1974] studied competitive adsorption of albumin, fibrinogen and γ-globulin onto different polymers (see also section IV). The experiments were carried out with an adsorption cell in which an air-solution interface was avoided. Generally, fibrinogen adsorbed faster onto any polymer than the other two proteins.

Several data from experiments cited here strongly suggest that *fibrinogen, adsorbed onto a surface, interacts with platelets.* However, KLINGS et al. [1972] found that platelets adhered onto glass coated with Polybrene (a basic, and low molecular weight polymeric quaternary ammonium salt), although this material did not adsorb fibrinogen out of plasma. It was suggested that platelets might adhere electrostatically onto this surface. On the other hand, KLINGS et al. [1972] observed that protamine sulfate films adsorbed fibrinogen and yet platelets did not adhere to the fibrinogen layer. This is even more difficult to understand. It was supposed that fibrinogen can perhaps be adsorbed in at least two (anti-genically reactive) different positions; one of them prevents the adhesion of platelets.

A mechanism for the interaction between platelets and adsorbed fibrinogen was proposed by KIM et al. [1974] and LEE and KIM [1974b]. In their model, glycosyltransferases located in the platelet membrane, complex with surface-adsorbed glycoproteins containing incomplete heterosaccharides. It was suggested that fibrinogen and γ-globulin would contain such incomplete chains; albumin contains no carbohydrate. This mechanism is similar to that proposed by *Jamieson* [1973] for the interaction of platelets with collagen (see section V.A.). *In vitro* experiments of LEE and KIM [1974b] showed that sugar-nucleotides could be transferred by platelet glycosyl transferases onto adsorbed fibrinogen and γ-globulin, but not onto albumin. The role of carbohydrate in platelet adhesion was also examined by measuring platelet adhesion (platelets in blood from anesthetized sheep) to tubes coated with proteins from which different carbohydrate residues were removed enzymatically. It was found that somewhat more platelets adhered to enzymatically treated fibrinogen and γ-globulin coatings than to the untreated coatings. This indicates that *platelet adhesion increases with an increasing amount of incomplete saccharide chains available on the protein surface.*

Generally, heparinized surfaces prevent thrombus formation for some time [LEININGER et al., 1972]. The experiments of VROMAN and ADAMS [1969a] and REMBAUM et al. [1973] cited earlier show that this nonthrombogenicity is, at least partially, due to the inhibition of platelet adhesion. It is doubtful whether heparin, which is bound to the surface, also acts as an anticoagulant (unless there is a slow elution of heparin molecules from the surface). In contrast, VROMAN [1970] found that *in vitro* heparin-

treated TDMAC-glass surfaces (TDMAC = tridodecylmethyl ammonium chloride) *adsorbed and activated factor XII to a significant degree.* KLINGS *et al.* [1972] reported that protamine sulfate glass surfaces, treated with heparin, inhibited platelet adhesion and enhanced coagulation of normal plasma. These are important findings; they show that certain foreign surfaces can activate the intrinsic coagulation and can inhibit platelet adhesion at the same time. Therefore, it seems likely that on any foreign surface, which is exposed to blood, activation of the intrinsic coagulation can occur independently of platelet adhesion. This is confirmed by experiments of MASON *et al.* [1972] in which deficiency of fibrinogen reduced platelet adhesion (see above) but certainly did not delay the activation of the intrinsic coagulation. The data of MASON *et al.* show that the intrinsic coagulation process was even somewhat stimulated. They suggest that this is due to the fact that factor XII has to compete with fibrinogen for adsorption onto a foreign surface, and that factor XII is more rapidly adsorbed and activated in the absence of fibrinogen than in its presence.

It was already mentioned that platelet adhesion to foreign surfaces also occurs when the intrinsic coagulation is inhibited because factor VIII or factor IX is lacking, or because heparin is present in the blood. In the case of factor XII-deficient blood, MASON *et al.* [1972] observed a small decrease of platelet adhesion onto glass, siliconized glass and teflon. Possibly, platelet-associated molecules of factor XII function as a cementing agent between the platelet and the foreign surface.

From the foregoing it must be concluded that *both platelet adhesion and activation of the intrinsic coagulation occur more or less independently when a foreign surface is exposed to blood.* Furthermore, the mechanisms of both processes seem to be quite different. Thus, a measured interaction between any material and platelets generally does not reflect the influence of this material on intrinsic coagulation. This may also be concluded from the literature. Many materials, mostly polymers, were tested for their blood compatibility by MASON *et al.* [1969]. In an *in vitro* system they determined the interaction with platelets in a *Stypven time system*, and the degree of activation of coagulation by measuring the *partial thromboplastin time*. For the materials tested, they did not find a straightforward correlation between these two parameters.

It must be assumed that *in later stages of blood clotting on a foreign surface, there will be a mutual interaction between the coagulation process and platelet aggregation.* When thrombin is formed on a material via the intrinsic pathway, platelet release will take place, followed by the formation of aggregates. Experiments of RODMAN and MASON [1972] do confirm this. They compared platelet adhesion and fibrin formation on commercially available Silastic rubber and silicone rubber without a silica filler (PDMS), using an *ex vivo* cell. After 10 min exposure to native blood, only few platelets adhered to both Silastic and PDMS, and fibrin was not observed. After 15 min exposure, however, large accumulations of erythrocytes, platelet aggregates, and fibrin were observed on the Silastic. On the other hand, after 20 min exposure of blood to PDMS no fibrin was formed and only few platelets in areas of surface irregularity were observed.

In their paper, RODMAN and MASON [1972] write: 'The presence of fibrin clotting in the 15 min specimen (read Silastic) indicated thrombin evolution. The absence of either fibrin clotting or significant platelet adhesion in the 10 min sample suggested that platelet adhesion and aggregation seen at 15 min may have resulted from the presence of thrombin rather than specific or primary interaction between platelets and polymer surface.' Further, they suppose that factor XII is activated on the silica particles present in Silastic, thus leading to the formation of thrombin.

It is interesting to emphasize that the differences between the two silicone rubbers, found with the *ex vivo* cell of RODMAN and MASON, agree with the findings of KOLOBOW et al. [1974]. These investigators found that *membrane lungs and perfusion circuits of silicone rubber free of silica filler, applied to sheep, were superior in blood compatibility to those of standard silicone rubber.*

When platelet adhesion and contraction occurs on a foreign material, ADP will be released. In section V.B. it was already mentioned that platelets which are stimulated by ADP can initiate the intrinsic coagulation by activating factor XII. It seems likely that this process also takes place on the platelet aggregates which are formed on a foreign surface, thus leading to the formation of thrombin followed by the deposition of fibrin.

There is not much known about the *mechanism of factor XII adsorption and activation*. Some general remarks on this subject have been stated in section IV. Possibly, factor XII is uncoiled when it is adsorbed onto a foreign surface, thereby showing the enzymatically active center necessary for the activation of factor XI [MARGOLIS, 1963; VROMAN, 1964; DONALDSON and RATNOFF, 1965]. The recent experiments of MCMILLIN and WALTON [1974] (mentioned in section IV), in which a special circular dichroism technique was used, showed that a drastic conformation change occurred when factor XII was adsorbed (from a solution in saline) onto quartz. The spectrum of a dried layer of adsorbed factor XII, however, markedly differed from that of factor XII in the 'wet' absorbed state. Therefore, it was concluded that water is important for conformation and may be essential for the activation of this factor.

It is well known that factor XII is rapidly adsorbed and activated onto negatively charged surfaces, such as glass and similar materials [MARGOLIS, 1957; RATNOFF and ROSENBLUM, 1958]. Factor XII is also activated on collagen [NIEWIAROWSKI et al., 1966; WILNER et al., 1968]. WILNER et al. [1968] found that addition of cationic proteins to collagen reduced the activation of factor XII. They also found that pepsin treatment of collagen, which removes predominantly negatively charged telopeptides, decreased the activation of factor XII. Moreover, esterification of collagen, in which 80–90% of the free carboxyl groups were converted, reduced the activation of factor XII by over 90%. Therefore, it was suggested that free carboxyl groups (of collagen) provided the negatively charged sites for factor XII activation.

We already mentioned that both heparin-treated TDMAC and heparin-treated protamine sulfate on glass surfaces activated factor XII to a significant degree. Many negatively charged aminosulfonate and carboxylate groups are present in heparin. We must assume that the heparinized surfaces also expose negative groups (although heparin is linked ionically to these materials, not all the functional groups of heparin will be complexed). On the other hand, factor XII is not activated on the normal vessel wall, which is also negatively charged [SAWYER and SRINIVASAN, 1973]. Moreover, many polymers are not negatively charged and coagulation occurs as well. SALYER et al. [1971] did not find a correlation between the clotting times of citrated blood (mainly coagulation times) and the negative charge of various materials consisting of ethylene-acrylic acid copolymers and ethylene-vinyl-alcohol-methacrylic acid terpolymers. OJA et al. [1969] found no significant adsorption of factor XII from plasma onto either ion-exchange resins bearing strongly anionic (sulfate, sulfonate) or weakly anionic (carboxylate) surfaces.

SALYER et al. [1971] also prepared several polyurethanes containing conductive fillers. Their results indicate that an optimum degree of conductivity (or resistivity) may be important for the intrinsic coagulation. In this connection it is interesting to mention that BRUCK [1973b] reported that a modified conducting aromatic polyimide showed long clotting times of human blood *in vitro*. He also found very light platelet aggregation, using platelet-rich plasma. BRUCK suggests that *electroconduction and semiconduction may be involved in the interaction of surfaces with factor XII and other plasma proteins, and platelets by an unknown mechanism.*

For the sake of optimism, we will end our brief and subjective part about the interaction of blood and foreign surfaces with the remark: if we know more about the interaction between proteins and foreign materials, we will certainly know more about platelet adhesion and activation of the intrinsic coagulation by such materials.

VI. Concluding Remarks

Up to now, the most promising surface chemistry parameter of biocompatibility is *the minimal interfacial free energy hypothesis*, but only for a limited range of small interfacial free energy water-solid systems. Generally, the blood-contact properties of biomaterials cannot be predicted by a single parameter.

It has been shown that competitive rather than single adsorption experiments are necessary to correlate protein adsorption (onto foreign materials) with blood compatibility. The possible relationship between protein adsorption and platelet adhesion and/or activation of the intrinsic coagulation is briefly mentioned in section IV and further elaborated on in section V.

When a foreign material is exposed to blood, both platelet adhesion and activation of the intrinsic coagulation occur more or less independently. It must be assumed that in the later stages of blood clotting on foreign surfaces, a mutual interaction between platelet aggregation and intrinsic coagulation will occur. Platelet adhesion is most probably promoted by fibrinogen and γ-globulin, adsorbed out of plasma or blood onto a foreign surface. On the other hand, albumin coatings do prevent platelet adhesion. Very little is known about the mechanism of factor XII adsorption and activation on a foreign surface.

A systematic study of the behavior of well-characterized materials with varying compositions toward protein adsorption, platelet adhesion and activation of intrinsic coagulation under both static and flow conditions is necessary to get more insight into the mechanism(s) of blood clotting on foreign materials. Using the minimal interfacial free energy hypothesis, *it is conceivable that certain hydrogel-type materials prevent both platelet adhesion and factor XII activation* (protein adsorption will be minimized).

Acknowledgements

The authors wish to thank Dr. D. BARGEMAN for his helpful suggestions. One of the authors (J.F.) appreciates the stimulating discussions with Dr. J. D. ANDRADE, Dr. S. W. KIM and Dr. D. J. LYMAN during his one year's stay at the University of Utah. We thank Mrs. H. M. J. HAMMINK-SIERS for typing the manuscript.

References

ADAM, N. K.: Physics and chemistry of surfaces (Dover, New York 1968).
ADAMSON, A. W.: Physical chemistry of surfaces (Interscience, New York 1967).
ANDRADE, J. D.: Coagulation resistant coatings. Diss. Abstr. *30:* 3614B (1970).
ANDRADE, J. D.: Interfacial phenomena and biomaterials. Med. Instrum. *7:* 110–120 (1973).
ANDRADE, J. D. and KIM, S. W.: Polymer surfaces, Prog. Chem. chem. Ind. (Korea) *12:* 11–26 (1972).
ANDRADE, J. D.; KUNITOMO, K.; WAGENEN, R. VAN; KASTIGIR, B.; GOUGH, D., and KOLFF, W. J.: Coated adsorbents for direct blood perfusion: hema/activated carbon. Trans. Am. Soc. artif. internal Organs *17:* 222–228 (1971).
ANDRADE, J. D.; LEE, H. B.; JHON, M. S.; KIM, S. W., and HIBBS, J. B.: Water as a biomaterial. Trans. Am. Soc. artif. internal Organs *19:* 1–7 (1973).
BAIER, R. E.: Role of surface energy in thrombogenesis. Bull N.Y. Acad. Med. *48:* 257–272 (1972).
BAIER, R. E.; LOEB, G. L., and WALLACE, G. T.: Role of an artificial boundary in modifying blood proteins. Fed. Proc. *30:* 1523–1528 (1971).
BISCHOFF, K. B.: Discussion of correlations of blood coagulation with surface properties of materials. J. biomed. Mater. Res. *2:* 89–93 (1968).
BRASH, J. L. and LYMAN, D. J.: Adsorption of plasma proteins in solution to uncharged, hydrophobic polymer surfaces. J. biomed. Mater. Res. *3:* 175–189 (1969).
BRASH, J. L. and LYMAN, D. J.: Adsorption of proteins and lipids to nonbiological surfaces; in HAIR Chemistry of biosurfaces, chapt. 5 Marcel Dekker, New York 1971).
BROPHY, J. H.: Thermodynamics of structures; in WULFF Structure and properties of materials, vol. 2, chapt. 3 (Wiley, New York 1964).
BRUCK, S. D.: Polymeric materials: current status of biocompatibility. Biomater. med. Devices artif. Organs. *1:* 79–98 (1973a).
BRUCK, S. D.: Intrinsic semiconduction, electronic conduction of polymers and blood compatibility. Nature, Lond. *243:* 416–417 (1973b).
BULL, H. B.: Adsorption of bovine serum albumin on glass. Biochim. biophys. Acta *19:* 464–471 (1956).
CHESSELLS, J. M. and BLACK, M. M.: Prosthetic materials and the blood; in BLACK Developments in biomedical engineering, pp. 18–32 (Chatto & Windus, London 1972).
DAMUS, P. S.; HICKS, M., and ROSENBERG, R. D.: Anticoagulant action of heparin. Nature, Lond. *246:* 355–357 (1973).
DAVIE, E. W. and RATNOFF, O. D.: Waterfall sequence for intrinsic blood clotting, Science *145:* 1310–1312 (1964).
DE BOER, J. H.: Atomic forces and adsorption. Adv. Colloid Sci. *3:* 1 (1950).
DETTRE, R. H. and JOHNSON, R. E., jr: Surface properties of polymers. I. The surface tensions of some molten polyethylenes. J. Colloid Interface Sci. *21:* 367–377 (1966).
DILLMAN, W. J. and MILLER, I. F.: On the adsorption of serum proteins on polymer membrane surfaces. J. Colloid Interface Sci. *44:* 221–241 (1973).
DONALDSON, V. H. and RATNOFF, O. D.: Hageman factor: alterations in physical properties during activation. Science *150:* 754–756 (1965).
DROST-HANSEN, W.: Chemistry and physics of interfaces. II, p. 203 (Am. Chemical Society, Washington 1971).
FALB, R. D.; TAKAHASHI, M. T.; GRODE, G. A., and LEININGER, R. I.: Studies on the

stability and protein adsorption characteristics of heparinized polymer surfaces, by radioisotope labeling techniques. J. biomed. Mater. Res. *1:* 239–251 (1967).
FOWKES, F. M.: Surface chemistry; in PATRICK Treatise on adhesion and adhesives, vol. 1, pp. 328 ff. (Marcel Dekker New York 1967).
FOWKES, F. M.: Attractive forces at interfaces. Ind. Engng. Chem. *56:* 40–52 (1964).
GIRIFALCO, L. A. and GOOD, R. J.: A theory for the estimation of surface and interfacial energies. I. Derivation and application to interfacial tension. J. phys. Chem. *61:* 904–909 (1957).
GOOD, R. J. and ELBING, E.: Generalization of theory of estimation of interfacial energies. Ind. Engng. Chem. *62:* 54–78 (1970).
GUEST, M. M.; BOND, T. B., and CRAWFORD, M. A.: A possible role of leucocytes in hemostasis; in DITZEL and LEWIS, Clinical aspects of microcirculation, part II, pp. 131–137 (Karger, Basel 1973).
HAANEN, C.: The nature of the factor IX activating principle; in HEMKER, LOELIGER and VELTKAMP Human blood coagulation, pp. 58–72 (Leiden University Press, Leiden 1969).
HALL, C. E.: Visualization of individual macromolecules with the electron microscope. Proc. natn. Acad. Sci. USA *42:* 801–806 (1956).
HEMKER, H. C.; and KAHN, M. J. P.: Reaction sequence of blood coagulation. Nature, Lond. *215:* 1201–1202 (1967).
HEMKER, H. C.: LOELIGER, E. A., and VELTKAMP, J. J. (eds.): Human blood coagulation (Leiden University Press, Leiden 1969).
HILDEBRAND, J. H. and SCOTT, R. L.: Solubility of non-electrolytes, pp. 402, 431. (Reinhold, New York 1950).
HOFFMAN, A. S.: Principles governing biomolecule interactions at foreign interfaces. J. biomed. Mater. Res. Symp. *5:* 77–83 (1974).
HOVIG, T.; JØRGENSEN, L.; PACKHAM, M. A., and MUSTARD, J. F.: Platelet adherence to fibrin and collagen. J. Lab. clin. Med. *71:* 29–40 (1968).
JAMIESON, G. A.: Role of glycoproteins in platelet function; in GERLACH, MOSER, DEUTSCH, and WILMANNS Erythrocytes, thrombocytes, leucocytes, pp. 209–210 (Thieme, Stuttgart 1973).
KAELBLE, D. H.: Dispersion-polar surface tension properties of organic solids. J. Adhes. *2:* 66–81 (1970).
KIM, S. W.; LEE, R. G.; OSTER, H.; LENTZ, D. J.; COLEMAN, D. L.; ANDRADE, J. D., and OLSEN, D.: Platelet adhesion to polymer surfaces. Trans. Am. Soc. artif. internal Organs *20:* 449–455 (1974).
KIPLING, J. J.: Adsorption from solutions of nonelectrolytes (Academic Press, New York 1965).
KLINGS, M.; ADAMS, A. L., and VROMAN, L.: Effects of protamine sulfate, polybrene and heparin on the behavior of plasma, plasma proteins, platelets and factor XII activity at interfaces. Thromb. Res. *1:* 507–526 (1972).
KOLOBOW, T.; STOOL, E. W.; WEATHERSBY, P. K.; PIERCE, J.; HAYANO, F., and SUAUDEAU, J.: Superior blood compatibility of silicone rubber free of silica filler in the membrane lung. Trans. Am. Soc. artif. internal Organs *20:* 269–277 (1974).
LEE, H. B.; SHIM, H. S., and ANDRADE, J. D.: Radiation grafting of synthetic hydrogels to inert polymer surfaces. I. Hydroxyethyl methacrylate. Polymer Preprints *13:* 729–735 (1972).
LEE, R. G.; ADAMSON, C., and KIM, S. W.: Competitive adsorption of plasma proteins onto polymer surfaces. Thromb. Res. *4:* 485–490 (1974).

LEE, R. G. and KIM, S. W.: Adsorption of proteins onto hydrophobic polymer surfaces: adsorption isotherms and kinetics. J. biomed. Mater. Res. *8:* 251–259 (1974a).

LEE, R. G. and KIM, S. W.: The role of carbohydrate in platelet adhesion to foreign surfaces. J. biomed. Mater. Res. *8:* 393–398 (1974b).

LEE, W. H., jr. and HAIRSTON, P.: Structural effects on blood proteins at the gas-blood interface. Fed. Proc. *30:* 1615–1622 (1971).

LEININGER, R. I.; COOPER, C, W.; FALB, R. D., and GRODE, G. A.: Nonthrombogenic plastic surfaces. Science *152:* 1625–1626 (1966).

LEININGER, R. I.; CROWLEY, J. P.; FALB, R. D., and GRODE, G. A.: Three years experience *in vivo* and *in vitro* with surfaces and devices treated by the heparin complex method. Trans. Am. Soc. artif. internal Organs *18:* 312–315 (1972).

LEMOS, P. C. P.; TOLOSA, E. M. C.; CAMARGO, E. E.; MAKSOUD, J. G.; LANGER, B.; STOLE, N. A. G.; BARRETO, T. M.; NASSER, A., and ZERBINI, E. J.: Adsorption of fibrinogen-^{125}I on stellite 21, study with polished and Teflon-covered surfaces. J. thorac. cardiovasc. Surg. *68:* 405–410 (1974).

LEONARD, E. F. and FRIEDMAN, L. I.: Mass transfer in biological systems. Chem. Engng. Prog. Symp. Ser. *66:* 59–71 (1970).

LYKLEMA, J. and NORDE, W.: Biopolymer adsorption with special reference to the serum albumin-polystyrene latex system. Croat. chem. Acta. *45:* 67–84 (1973).

LYMAN, D. J.; BRASH, J. L.; CHAIKIN, S. W.; KLEIN, K. G., and CARINI, M.: The effect of chemical structure and surface properties of synthetic polymers on the coagulation of blood. II. Protein and platelet interaction with polymer surfaces. Trans. Am. Soc. artif. internal Organs. *14:* 250–255 (1968).

LYMAN, D. J. and KIM, S. W.: Interactions at the blood-polymer interface. Fed. Proc. *30:* 1658–1662 (1971).

LYMAN, D. J.; KLEIN, K. G.; BRASH, J. L., and FRITZINGER, B. K.: Interaction of platelets with polymer surfaces. I. Uncharged hydrophobic polymer surfaces. Thromb. Diath. haemorrh. *23:* 120–128 (1970a).

LYMAN, D. J.; KLEIN, K. G.; BRASH, J. J.; FRITZINGER, B. K.; ANDRADE, J. D., and BONOMO, F. S.: Platelet interaction with protein-coated surfaces: an approach to thromboresistant surfaces. Thromb. Diath. haemorrh., suppl. 42, pp. 109–114 (1970b).

LYMAN, D. J.; MUIR, W. M., and LEE, I. J.: Effect of chemical structure and surface properties of polymers on the coagulation of blood. I. Surface free energy effects. Trans. Am. Soc. artif. internal Organs *11:* 301–306 (1965).

MACFARLANE, R. G.: An enzyme cascade in the blood clotting mechanism, and its function as a biochemical amplifier. Nature, Lond. *202:* 498–499 (1964).

MARGOLIS, J.: Initiation of blood coagulation by glass and related surfaces, J. Physiol., Lond. *137:* 95–109 (1957).

MARGOLIS, J.: The interrelationship of coagulation of plasma and release of peptides. Ann. N.Y. Acad. Sci. *104:* 133–145 (1963).

MASON, R. G.; SCARBOROUGH, D. E.; SABA, S. R.; BRINKHOUS, K. M; IKENBERRY, L. D.; KEARNEY, J. J., and CLARK, H. G.: Thrombogenicity of some biomedical materials: platelet-interface reactions. J. biomed. Mater. Res. *3:* 615–644 (1969).

MASON, R. G.; SHERMER, R. W., and RODMAN, N. F.: Reactions of blood with nonbiological surfaces. Am. J. Path. *69:* 271–288 (1972).

MASON, R. G.; SHERMER, R. W.; ZUCKER, W. H., and BRINKHOUS, K. M.: Platelets and plasma protein reactions with foreign surfaces; in GERLACH, MOSER, DEUTSCH, and WILMANNS Erythrocytes, thrombocytes, leucocytes, pp. 263–266 (Thieme, Stuttgart 1973).

MCMILLIN, C. R. and WALTON, A. G.: A circular dichroism technique for the study of adsorbed protein structure. J. Colloid Interface Sci. *48:* 345–349 (1974).

McRitchie, F.: The adsorption of protein at the solid/liquid interface. J. Colloid Interface Sci. *38:* 484–488. (1972).
Morrissey, B. W. and Stromberg, R. R.: The conformation of adsorbed blood proteins by infrared bound fraction measurements. J. Colloid Interface Sci. *46:* 152–164 (1974).
Niewiarowski, S.; Stuart, R. K., and Thomas, D. P.: Activation of intravascular coagulation by collagen. Proc. Soc. exp. Biol. Med. *123:* 196–199 (1966).
Oja, P. D.; Holmes, G. W; Perkins, H. A., and Love, J.: Specific coagulation factor adsorption as related to functional groups on surfaces. J. biomed. Mater. Res. *3:* 165–174 (1969).
Owens, D. K. and Wendt, R. C.: Estimation of the surface free energy of polymers. J. appl. Polymer Sci. *13:* 1741–1747 (1969).
Packham, M. A.; Evans, G.; Glynn, M. F., and Mustard, J. F.: The effect of plasma proteins on the interaction of platelets with glass surfaces. J. Lab. clin. Med. *73:* 686–697 (1969).
Patterson, D. and Rastogi, A. K.: The surface tension of polyatomic liquids and the principle of the corresponding states. J. phys. Chem. *74:* 1067–1071 (1970).
Patterson, D. D. and Siow, K. S.: Prediction of surface tensions of liquid polymers. Macromolecules *4:* 26–30 (1971).
Patterson, H. T.; Hu, K. H., and Grindstaff, T. H.: Measurements of interfacial and surface tensions in polymer melts. J. Polymer Sci., C *34:* 31–43 (1971).
Prigogine, J.; Bellemans, A., and Mathot, V.: The molecular theory of solutions, chapt. 16 (North Holland/Interscience, Amsterdam/New York 1957).
Ratnoff, O. D. and Rosenblum, J. M.: Role of Hageman factor in the initiation of clotting by glass. Am. J. Med. *25:* 160–168 (1958).
Rembaum, A.; Yen, S. P. S.; Ingram, M.; Newton, J. F., and Hu, C. L.: Platelet adhesion to heparin-bonded and heparin free surfaces. Biomater. med. Devices artif. Organs. *1:* 99–119 (1973).
Rodman, N. F. and Mason, R. G.: Blood-foreign surface interaction. Thromb. Diath. haemorrh., suppl. 42, pp. 61–71 (1972).
Roe, R. J.: Parachor and surface tension of amorphous polymers. J. phys. Chem. *69:* 2809–2810 (1965).
Roe, R. J.: Hole theory of surface tension of polymer liquids. Proc. natn. Acad. Sci. USA *56:* 819–824 (1966).
Roe, R. J.; Surface tension of polymer liquids. J. phys. Chem. *72:* 2013–2017 (1968).
Safonov, G. P. and Entelis, S. G.: Surface properties and parachor of liquid polyetherdiols. Vȳsokomolek. Soedin. ser. A *9:* 1909–1913 (1967).
Sakai, T.: Surface tension of polyethylene melt. Polymer. *6:* 659–661 (1965).
Salyer, I. O.; Blardinelli, A. J.; Ball, G. L., III; Weesner, W. E.; Gott, V. L.; Ramos, M. D., and Furuse, A.: New blood-compatible polymers for artificial heart applications. J. biomed. Mater. Res. Symp. *1:* 105–127 (1971).
Salzman, E. W.; Merrill, E. W.; Binder, A.; Wolf, C. F. W.; Ashford, T. P., and Austen, W. G.: Protein-platelet interactions on heparinized surfaces. J. biomed. Mater. Res. *3:* 69–81 (1969).
Sawyer, P. N. and Srinivasan, S.: The role of surface phenomena in intravascular thrombosis; in Ditzel and Lewis Clinical aspects of microcirculation, part II, pp. 106–119 (Karger, Basel 1973).
Schonhorn, H.: Surface free energy of polymers. J. phys. Chem. *69:* 1084–1085 (1965).
Schonhorn, H.: Heterogeneous nucleation of polymer melts. I. Influence of substrates on wettability. J. Polymer Sci. *5B:* 919–924 (1967).

SCHONHORN, H.: Heterogeneous nucleation of polymer melts. II. Effect of substrate on morphology and wettability. Macromolecules *1:* 145–151 (1968).

SCHONHORN, H. and RYAN, F. W.: Wettability of polyethylene single crystal aggregates, J. phys. Chem. *70:* 3811–3815 (1966).

SCHONHORN, H. and RYAN, F. W.: Adhesion of polytetrafluoroethylene. J. Adhes. *1:* 43–47 (1969).

SHIONOYA, T. and ROWNTREE, L. G.: Studies in experimental extracorporeal thrombosis. J. exp. Med. *46:* 7, 13, 19, 957, 963 (1927).

SPAET, T. H. and LEJNIEKS, I.: A technique for the estimation of platelet-collagen adhesion. Proc. Soc. exp. Biol. Med. *132:* 1038–1041 (1969).

STEWART, C. W. and FRANKENBERG, C. A. VON: Significant structures theory of the surface tension of polyethylene, J. Polymer Sci., A-2, *6:* 1686–1688 (1968).

STONER, G. E.; SRINIVASAN, S., and GILEADI, E.: Adsorption inhibition as a mechanism for the antithrombogenic activity of some drugs. I. Competitive adsorption of fibrinogen and heparin on mica, J. phys. Chem. *75:* 2107–2111 (1971).

SUDGEN, S.: The variation of surface tension with temperature and some related functions. J. chem. Soc. *125:* 32–41 (1924).

VALENTINE, R. C.: The shape of protein molecules suggested by electron microscopy. Nature, Lond. *184:* 1838–1841 (1959).

VROMAN, L.: Effects of hydrophobic surfaces upon blood coagulation. Thromb. Diath. haemorrh. *10:* 455–493 (1964).

VROMAN, L.: Surface activity in blood coagulation; in SEEGERS Blood clotting enzymology, pp. 279–322 (Academic Press, New York 1967).

VROMAN, L.: Identification of proteins adsorbed out of plasma onto biomaterials. Prog. Rep. PB-193886, NIAMD (National Technical Information Service, Springfield, Va. 1970).

VROMAN, L.: Identity and significance to clotting and platelet adhesion, of proteins adsorbed out of plasma onto artificial kidney membrane materials. Prog. Rep. PB-204601, NIAMD (National Technical Information Service, Springfield, Va. 1971).

VROMAN, L. and ADAMS, A. L.: Identification of rapid changes at plasma-solid interfaces, J. biomed. Mater. Res. *3:* 43–67 (1969a).

VROMAN, L. and ADAMS, A. L.: Effect of heparin on reactions at aminated polymer-blood interfaces. J. Colloid Interface Sci. *31:* 188–195 (1969b).

VROMAN, L.; ADAMS, A. L., and KLINGS, M.: Interaction among human blood proteins at interfaces. Fed. Proc. *30:* 1494–1502 (1971).

WALSH, P. N.: The role of platelets in the contact phase of blood coagulation. Br. J. Haemat. *22:* 237–254 (1972a).

WALSH, P. N.: The effects of collagen and kaolin on the intrinsic coagulant activity of platelets. Evidence for an alternative pathway in intrinsic coagulation not requiring factor XII. Br. J. Haemat. *22:* 393–405 (1972b).

WALSH, P. N.: Platelet coagulant activities and hemostasis: a hypothesis, Blood *43:* 597–605 (1974).

WALSH, P. N. and BIGGS, R.: The role of platelets in intrinsic factor Xa formation. Br. J. Haemat. *22:* 743–760 (1974).

WEISS, H. J.; Inhibition of platelet-induced thrombus formation. Schweiz. med. Wschr. *104:* 114–119 (1974).

WICHTERLE, O. and LIM, D.: Hydrophilic gels for biological use. Nature, Lond. *185:* 117–118 (1960).

WILLIAMS, W. J.: Introduction to the plasma coagulation factors; in WILLIAMS, BEUTLER, ERSLEV and RUNDLES Hematology, pp. 1055–1056 (McGraw-Hill, New York 1972).

WILLIAMS, W. J.; BEUTLER, E.; ERSLEV, A. J., and RUNDLES, R. W.: Hematology (McGraw-Hill, New York 1972).

WILNER, G. D.; NOSSEL, H. L., and LeROY, E. C.: Activation of Hageman factor by collagen, J. clin. Invest. 47: 2608–2615 (1968).

WU, S.: Estimation of the critical surface tension for polymers from molecular constitution by a modified Hildebrand-Scott equation. J. phys. Chem. 72: 3332–3334 (1968).

WU, S.: Surface and interfacial tensions of polymer melts. I. Polyethylene, polyisobutylene, and polyvinyl acetate. J. Colloid Interface Sci. 31: 153–161 (1969).

WU, S.: Surface and interfacial tensions of polymer melts. II. Poly(methyl methacrylate), poly(n-butyl methacrylate), and polystyrene. J. phys. Chem. 74: 632–638 (1970).

WU, S.: Polar and nonpolar interactions in adhesion. J. Adhes. 5: 39–55 (1973).

WU, S.: Interfacial and surface tensions of polymers. J. macromol. Sci. C 10(1): 1–73 (1974).

ZISMAN, W. A.: Relation of the equilibrium contact angle to liquid and solid constitution; in Contact angle, wettability, and adhesion. Adv. Chem. Ser. vol. 43, pp. 1–51 (Am. Chemical Society, Washington 1964).

ZUCKER, M. B.: Platelet function; in WILLIAMS, BEUTLER, ERSLEV, and RUNDLES Hematology, pp. 1014–1022 (McGraw-Hill, New York 1972).

ZUCKER, M. B. and VROMAN, L.: Platelet adhesion induced by fibrinogen adsorbed onto glass. Proc. Soc. exp. Biol. Med. 131: 318–320 (1969).

Dr. J. FEIJEN, Polymer Division, Department of Chemical Technology, Twente University of Technology, *Enschede* (The Netherlands)

Blood-Surface Interactions as a Basis for Selection of Blood-Compatible Cardiovascular Implantable Materials

S. SRINIVASAN, N. RAMASAMY, B. STANCZEWSKI and P. N. SAWYER

Electrochemical and Biophysical Laboratories, Departments of Surgery and Surgical Research, State University of New York, Brooklyn, N.Y.

Contents

Abstract 134
Glossary 134
   I. Introduction 135
  II. Some Essential Criteria for the Selection of Blood-Compatible Materials ... 135
 III. Correlations Between Interfacial Electrochemical Properties and Blood Compatibility of Materials 136
     A. Conducting Materials 136
     B. Insulator Materials 143
 IV. Some Recent Techniques for Improving Blood Compatibility 148
     A. Biolized Materials 148
     B. Linear or Cross-Linked Homo or Block Polymers 149
     C. Surface Pretreatment of Metallic Materials 149
  V. Blood-Surface Interactions – Mechanism Studies 150
     A. General 150
     B. Effect of Metals on the Platelet Release Reaction 151
     C. Electron Microscopic and Optical Studies of Surfaces Exposed to Biologic Fluids 152
     D. Electrochemical Reactions of Blood Coagulation Factors 154
     E. Adsorption and Adsorption Inhibition 158
 VI. Conclusions 159
Acknowledgements 160
References 160

Abstract

Thrombosis on the blood vessel wall or on prosthetic materials is one of the most serious cardiovascular problems. Thrombosis occurs at a solid-electrolyte interface and is markedly influenced by the electrochemical characteristics across the interface. In the present chapter is presented a summary of the results on the correlation between the electrochemical and antithrombogenic characteristics of materials.

Conducting materials, registering potentials below +100 mV versus Normal Hydrogen Electrode (NHE) in blood, tend to be nonthrombogenic. The corrodible metals belong to this class. Noble metals are thrombogenic. However, by maintaining them at negative potentials in blood, versus NHE, they can be made nonthrombogenic. Metals and alloys, presently used in production of cardiac valves, remain passive in blood and physiologic saline. In practical cardiac prostheses, the surface contamination appears to play a major role in thrombogenesis.

Polymeric materials exhibiting negative zeta potentials generally tend to be nonthrombogenic. Chemical treatment of insulator materials to introduce negatively charged groups, such as sulfonate and carboxylate, improves the blood compatibility. Cross-linking of heparin on surfaces and biolization of materials are new approaches to preparing stable nonthrombogenic materials.

Blood-surface interactions, investigated at a molecular level, are also briefly outlined. Blood coagulation factors take part in adsorption and charge transfer reactions across metal-solution interfaces.

Glossary

Anticoagulant: A drug which prevents the formation of a coagulum.

Antithrombotic: A drug which prevents the formation of a thrombus.

Charge transfer or electron transfer reactions: An interfacial reaction where there is transfer of charge across the interface, usually electrons.

Coagulation: The formation of a solid from blood (coagulum).

Coagulation cascade: Sequential reaction of clotting factors to form a coagulum.

Coagulum: The solid formed from blood via the activation of the coagulation cascade. composed primarily of polymerized fibrin.

Electrokinetic phenomena: These arise due to relative movement of ions along the shear plane in the electric double layer across an insulator solution interface. Electrophoresis, streaming potential, and electroosmosis come under this category. Measurement of any of these electrokinetic parameters of blood cells and cellular fragments gives an indication of their surface charge. The measured parameters are related to the zeta potential.

Hemostasis: The cessation of bleeding.

Interfacial reaction: A reaction that takes place at the junction of two phases; blood vessel ↔ blood interface.

Platelet aggregation: The specific binding together of platelets in the formation of a 'plug' as an early stage in hemostasis.

Platelets: Cell fragments involved in primary hemostasis.

Primary hemostasis: The initial cessation of bleeding due to adhesion and aggregation of platelets at the site of injury.

Prosthesis: Something artificial which replaces a body structure or part.

White blood cells: Blood cells involved primarily in host defense (leukocytes).

Zeta potentials: The potential drop across the mobile part of the double layer.

I. Introduction

Atherosclerosis, acute thrombotic occlusions and embolization represent the most frequently encountered processes which result in fatalities due to cardiovascular disease. There are more than 5 million Americans who suffer from cardiovascular disease. Efforts are being made continuously to find methods of preventing and treating the disease. One method of treatment utilizes artificial prostheses (e.g., cardiac valves and bypass grafts) in place of the natural ones. Here, it becomes necessary to find blood-compatible materials for use in the fabrication of cardiovascular prostheses. Some of the results of our basic and clinical investigations, carried out in our search for blood-compatible materials, are presented in this article.

II. Some Essential Criteria for the Selection of Blood-Compatible Materials

Thrombosis is an interfacial reaction taking place at the blood vessel wall or prosthetic material-blood interface. The kinetics of all interfacial reactions are dependent on the electric potential drop across their double layers. The effects of electrochemical factors on blood coagulation dates back to 1824 when it was reported that, *in vitro*, blood is deposited on the positive electrode but not on the negative electrode [SCUDAMORE, 1824]. The application of electric currents in the therapy of thrombosis, aneurysms and hemostasis was also carried out in the 19th century [POORE, 1876; SKENE, 1899]. In 1928, it was proposed that materials with high negative zeta potentials favor adsorption of positively charged blood elements, thereby concentrating the necessary elements for the initiation of blood coagulation [GORTNER and BRIGGS, 1928]. Plasma fractions and chemicals which affect the coagulation mechanism create uniquely different surface charge effects on materials [WOOD *et al.*, 1950]. There was no direct correlation between zeta potentials and clotting activities of some materials, according to some workers [LEININGER *et al.*, 1964]. However, in a later work by this group, it was concluded that zeta potential measurements offer a convenient way of studying adsorption of blood components on plastics and that this study may lead to an understanding of the interactions of blood with plastics [LEININGER *et al.*, 1966]. Another group has supported our hypothesis that positively charged surfaces are prone to thrombosis [MILLIGAN *et al.*, 1966]. Recently, this group concluded that streaming potential techniques may serve as a quality control in the preparation of antithrombogenic materials [EDMARK *et al.*, 1970].

The surface energy of a material has been suggested as an important criterion in the selection of blood-compatible materials [BAIER and DUTTON, 1969; LYMAN et al., 1969]. The related parameters, surface tension and contact angle, have also been determined for some materials and correlated with their antithrombogenic activity. Platelet adhesion decreases with decreasing critical surface tension of materials. Lower energy surfaces are less thrombogenic. The effects of several functional groups on polymer surfaces (hydroxyl, sulfate, sulfhydryl, carboxylic, amino, sulfamido and quarternary ammonium salt) and on clotting factor interactions have been determined [GRODE et al., 1969]. Electrets with a negative surface charge tend to be antithrombogenic, and the extent of platelet adhesion on them is less than on nonpolarized polymers [MURPHY et al., 1969].

Polymer surfaces precoated with a plasma protein monolayer, particularly albumin, show a marked reduction in platelet adhesion [LYMAN et al., 1969]. Thrombus deposition on various surfaces has been observed with varying velocities of flow. Platelets are usually the first formed elements to attach to surfaces. In all regions where thrombi form, leukocytes are also deposited on the surfaces. Surfaces which show least leukocyte adhesion appear to be the most antithrombogenic [PETSCHEK and MADRAS, 1969].

During the last 15 years, correlations have been drawn between the electrochemical and antithrombogenic characteristics of several materials, and this will be dealt with in the next section [SAWYER and SRINIVASAN, 1967; SRINIVASAN and SAWYER, 1969, 1970; SRINIVASAN et al., 1973].

III. Correlations Between Interfacial Electrochemical Properties and Blood Compatibility of Materials

A. Conducting Materials

A knowledge of the rest potentials of materials in blood or other electrolytes with similar ionic composition is essential in predicting their antithrombogenic or thrombogenic behavior. Materials with negative potentials, versus Normal Hydrogen Electrode (NHE) in physiologic electrolytes tend to be nonthrombogenic, while materials registering positive potentials are thrombogenic [CHOPRA et al., 1967].

The potential of the material is measured in the desired electrolyte with respect to a reference electrode. To remove surface impurities, the materials are first treated with nitric or hydrochloric acid.

Three types of studies have been carried out to determine the implantation behavior of materials: (a) Wires of some metals have been inserted through side branches in canine carotid and femoral arteries and left in position for 30–90 min. They were then carefully removed and examined for thrombus deposits. (b) Tubes of several metals and alloys have been implanted in the canine descending thoracic aorta (DTA) or thoracic inferior vena cava (TIVC) in short-term (2 h) and long-term (more than 2 weeks) studies. (c) Implantation of valves of materials in the aortic, tricuspid or mitral annuli of calves. Detailed results on the potentials of metals and alloys in blood and on their behavior on implantation as tubes and valves are found in table I.

Table I. Results of implantation of valves and tubes, fabricated with conducting materials, in animal heart aorta and thoracic inferior vena cava

Metal or alloy	Rest potential in blood, mV/NHE			Thrombogenic behavior	
	before pre-treatment	after 10 N HCl cleaning	post mortem	tube % occlusion 0%..........100%	valve % occlusion 0%..........100%
Mg	−1500 to −1400	−		○ ═══	⟨↕⟩ ○
Mg (Mg, Al, Zn)	−1110 to −420	−1350 to −500		○ ═══	⟨↕⟩ ○
Al	−750 to −450	−780 to −540	−560 to −290	○ ═══	⟨↕⟩ ○
Al (Al, Cu)	−420 to −390	−530 to −420	−480 to −280	○ ═══	⟨↕⟩ ○
Al	−760 to 630	electro-polished	−780 to −730	○ ═══	⟨↕⟩ ○
Al	−485 to −775	glow discharge	−560 to −460	○ ═══	
Ti	−250 to +10	−350 to −200	−190 to −130		⟨↔⟩ ○
Cd	−250 to −210	−320 to −280	−		⟨↕⟩ ○
Co	−240 to −160	−300 to −260	−		⟨↕⟩ ○
Ni	−250 to −230	−300 to −280	−		⟨↕⟩ ○

Table I (continued)

Metal or alloy	Rest potential in blood, mV/NHE			Thrombogenic behavior	
	before pre-treatment	after 10 N HCl cleaning	post mortem	tube % occlusion 0%..........100%	valve % occlusion 0%..........100%
Stellite-21[1] Co, Cr, Mo, Fe, Ni, Mn, Si, C	−260 to −30	−285 to −160	−710 to +700	●━━	⋒ ○
Stellite-21[2]	−260 to +300	−280 to −160	−20 to +130		⋒ ○
Stellite-21[2] covered with Teflon fabric	−180 to +20	not pretreated	−470 to +110		⋒ ○
Vitallium-P 401 C Co, Cr, Mo, Ni, Mn, Si, Fe, C	+15 to +175	−210 to +270	−60 to +190	○━━	
Vitallium-P 103 C Co, Gr, Wi, Ni, Fe, Mn, Si, C	+70 to +150	−180 to +160	+60 to +220	○━━	
Haynes-25 Co, Gr, W, Ni, Fe, Mn, Si, C	−40 to +360	−180 to −80	−100 to +190		⋒ ○
Inconel Ni, Cr, Fe, Ta, Si, Nb, Mn, C, Co	+30 to +70	−120 to +10	+185 to +90		⋒ ●
Hastalloy 'B' Ni, Mo, Fe, Mn, Si, V, Cr, C	−110 to +185	−180 to −160	−180 to +140		⋒ ◐
Stainless steel-304 Fe, Cr, Ni, Mn, C, Si, P, Si	+110 to +190	−160 to −140	+110 to +130		⋒ ○
Stainless steel-309 Fe, Cr, Ni, Mn, Si, P, C, S	+90 to +110	−160 to −110	+10 to +230		⋒ ○
N-155 Fe, Cr, Co, Ni, Mo, W, Mn, Si, C	+70 to +120	−310 to −130	+30 to +200		⋒ ●
HCR	+110 to +210	−130 to −100	+130 to +170	●━━	

[1] Arwood.
[2] Starr-Edwards.

Table I (continued)

Metal or alloy	Rest potential in blood, mV/NHE		Thrombogenic behavior	
	before pre-treatment	post mortem	tube % occlusion 0%..........100%	valve % occlusion 0%..........100%
Porous Carbon[a]	+300 to +320	not treated +300 to +310	~10%	
Vitreous Carbon[a]	+280 to +970	+300 to +90	~10%	~40%
Porous Carbon[b]	+280 to +220	+230 to +180	~10%	
Pyrolytic Carbon[b]	+320 to +570	+310 to −280	~10%	
Bioelectric Polyurethane[c]	−160 to −140	−160 to −10		~30%
Electrolour	+320 to +360	−460 to −410	~10%	
*Teflon-*coated Stellite-21 valves untreated	−	−		~40%
Sulfonated *Teflon* (coated Stellite-21) 1.3–0.6 mol/cm²	−140 to −80	−260 to +60		~40%
Carboxylated *Teflon* (coated Stellite-21) 0.53 mol/cm²	−10 to +40	+210 to +180		~40%
Untreated *Polypropylene*	−	−		0%
Sulfonated *Polypropylene*	−	−		~20%

[a] Le Carbone.
[b] Gulf General Atomic.
[c] Goodyear Research Lab.

Table I (continued)

Metal or alloy	Rest potential in blood, mV/NHE			Thrombogenic behavior	
	before pre-treatment	after 10 N HCl cleaning	post mortem	tube % occlusion 0%..........100%	valve % occlusion 0%..........100%
Cu cathodically polarized	+90 to +150	−10 to −160	−20 to +40	(minimal occlusion)	
Cu ac polarized 100,000/sec	+90 to +180	+40 to +90	+20 to +80	(minimal occlusion)	
Cu glow discharge	+120 to +130	glow discharge	+160 to +165	(partial occlusion)	
Cu	+90 to +350	−60 to +140	+40 to +70	(partial occlusion)	(high occlusion)
Ag	+800	+280	—		(high occlusion)
Pt	+860	+320	—		(high occlusion)
Au	+1500	+430			(high occlusion)

Corrodible metals and alloys establish negative potentials in blood, versus NHE, and tend to be antithrombogenic. The more noble metals record stable positive potentials and are thrombogenic. In some of our studies, the potentials of implanted valves were followed *in vivo* as a function of time of implantation [LEE *et al.*, 1972]. Valves fabricated with aluminum maintain negative potentials *in vivo*, which decrease with time till some stable value is reached. Some aluminum valves have been found to remain thrombus-free for over 24 months.

Titanium has been used in the manufacture of cardiac valves. Our animal studies show that with increasing time of implantation, a stable positive, but numerically small, potential is established. A valve removed 12 months after implantation is shown in figure 1 to be free of thrombus deposits. Certain stainless steels (304, 309) used as orthopedic implants, however, develop positive potentials in blood and are thrombogenic. Copper and silver are markedly thrombogenic materials. Some Co-Cr alloys, Stellite-21 (fig. 2), vitallium are promising antithrombogenic materials (table I).

TITANIUM (SMELLOF-CUTTER)

INCONEL WIRE
UNTREATED
IMPLANTED: 12 MONTHS
CAUSE OF DEATH: SACRIFICE
CONDITION OF VALVE: NO THROMBOSIS

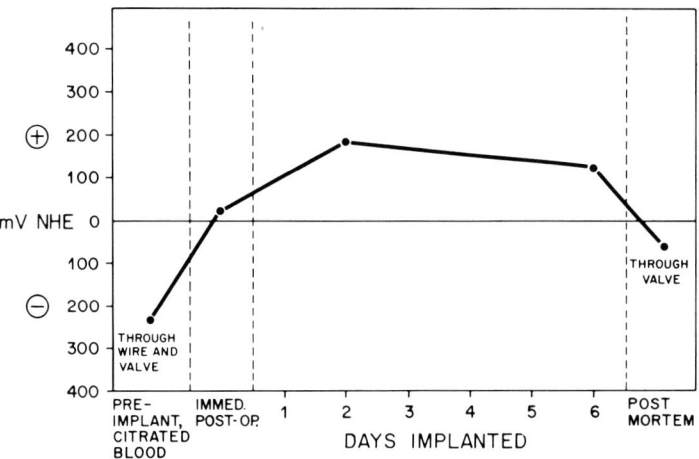

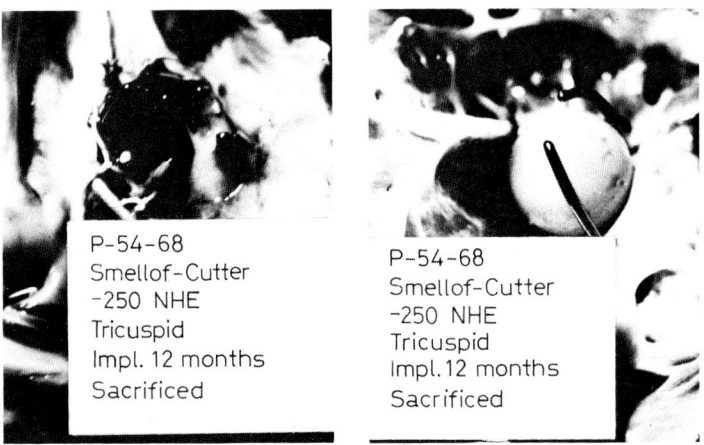

Fig. 1. The potential-time behavior of a titanium valve implanted in the tricuspid annulus of a calf for 12 months. The valve is free of thrombus deposits.

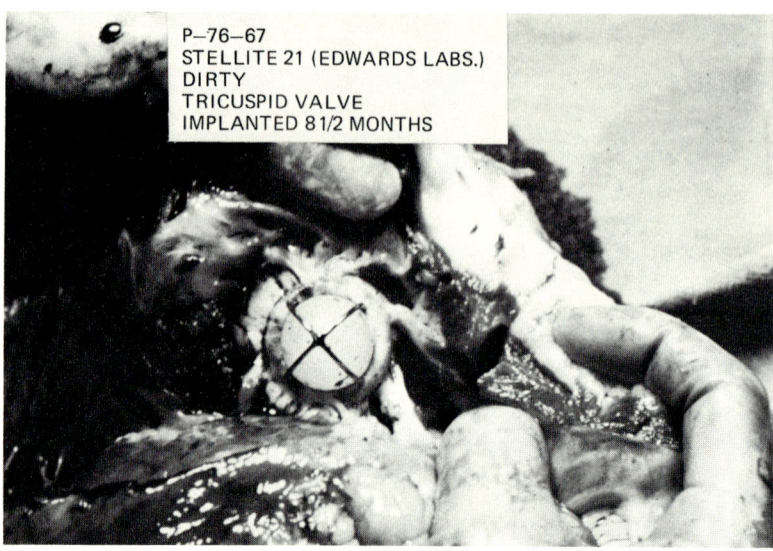

Fig. 2. Photograph of a thrombus-free stellite valve, removed after $8\frac{1}{2}$ months of implantation in the tricuspid annulus of a calf.

Fig. 3. A bioelectric polyurethane valve implanted for 489 days in the tricuspid annulus of a calf is free of thrombus deposits.

A novel electrochemical method of inhibiting thrombus deposition on a noble metal like copper is by maintaining the surface at a cathodic potential (vs NHE) using a suitable polarizing circuit [GILEADI et al., 1972]. Thus cathodically polarized copper tubes implanted in the canine TIVC were found to be patent after 2 weeks while the unpolarized copper implants were occluded within a day.

Polished pyrolytic carbon is a promising blood-compatible material. It has some advantages over the antithrombogenic metals because of its inertness in physiologic fluids. Polyurethane made conducting with carbon, (bioelectric polyurethane) shows good blood-compatible behavior (fig. 3).

B. Insulator Materials

Using electrokinetic techniques, it is possible to determine the zeta potentials across polymer-solution interfaces. When studying different materials in the same electrolyte, the order of increasing or decreasing zeta potentials is generally the same as the corresponding order of interface charge densities. By measuring the streaming potential developed in a tube of an insulator material during flow of electrolyte, one is able to deduce the zeta potentials. With porous materials, electroosmosis techniques yield similar data.

In our laboratories, correlations have been drawn between the net surface charge and antithrombogenic characteristics of a wide variety of insulator materials (table II). Positively charged materials are thrombogenic. However, not all negatively charged surfaces are nonthrombogenic. This is probably related to the heterogeneity of the surface charge. The introduction of sulfonate or carboxylate groups on the surface of Teflon markedly increases the magnitude of the zeta potential (since Teflon is negatively charged, these groups make Teflon even more negative) and also improves the blood compatibility. Hydroxyl groups have a smaller effect. Several groups of workers have attempted to bond heparin to plastic surfaces (silicone rubber and epoxy). In most cases, the heparin is bonded ionically to the materials. This is not stable in electrolytes and hence there is elution of heparin from heparinized vascular or heart valve prostheses implanted in animals. Covalent bonding of heparin to surfaces has not been very successful. Cross linking of ionically bonded heparin to surfaces, using gluteraldehyde, appears to inhibit heparin elution from surfaces [LAGERGREN and ERICKSSON, 1971].

Electrets with a negative surface charge tend to be antithrombogenic. Electrification does not increase the negative charge density but results in a more uniform distribution of charges. Fluoro-silicone, dimethyl-silicone

Table II. Results of implantation of tubes, fabricated with insulator materials, in canine thoracic aorta and/or thoracic inferior vena cava

Material	Zeta potential in Krebs solution mV	Thrombogenic behavior % occlusion 0% 100%
Untreated *Epoxy*	−5.3	
Epoxy with non-ionic detergent	−4.8	
Heparinized *Epoxy* 3%	−15.7	
Heparinized *Epoxy* 7%	−26.5	
Neutral *Ioplex*		
(0.5 Cl⁻) *Ioplex* cationic	+5.6	
(0.5 H⁺) *Ioplex* anionic	−7.8	
Hydrogel		
Untreated *Dacron*	−	
Acrylate *Dacron* 1.15–2.5 mol/cm²	−	
Crotonate *Dacron* 1.1 mol/cm²	−	
PVC *Dacron* 1.15 mol/cm²	−	
Ethylene sulfonic acid *Dacron* 2.5–0.2 mol/cm²	−	

Table II (continued)

Material	Zeta potential in Krebs solution mV	Thrombogenic behavior % occlusion 0%............100%
Untreated Teflon	−9.2	
Sulfonated Teflon 1.23 mol/cm^2	−31.5	
Sulfonated Teflon 1.02 mol/cm^2	−18.4	
Sulfonated Teflon 0.2–0.6 mol/cm^2	−26.5	
Sulfonated Teflon 0.02 mol/cm^2	−13.4	
Sulfonated Poly (vinyl amine) Teflon	−16.9	
Poly N-vinyl succinamic acid Teflon 0.43 mol/cm^2	−23.4	
Hydroxylated Teflon	−29.0	
Poly (maleic acid) Teflon 0.96 mol/cm^2	−27.3	
Carboxylated Teflon 0.75 mol/cm^2	−31.4	
Carboxylated Teflon 0.40 mol/cm^2	−30.3	
Carboxylated Teflon 0.30 mol/cm^2	−27.4	
Thermally treated Teflon	−11.0	

Table II (continued)

Material	Zeta potential in Krebs solution mV	Thrombogenic behavior % occlusion 0% 100%
Electrically treated *Teflon*	−12.0	
Porous *Teflon*[1]	—	
Electrets	−6.3	
Untreated *Silicone*	−3.2	
Heparinized *Silicone*	−3.4	
TDMAC *Silicone*	+4.8	
Silicone[2]	—	
Dimethyl *Silicone* homopolymer	−6.4	
Fluoro *Silicone* homopolymer	−5.8	
Fluoro *Silicone* dimethyl copolymer	−8.5	
Fluoro *Silicone* dimethyl duromer	−8.5	
PVC	−9.3 to −7.4	
Polyurethane		

[1] Gore-tex.
[2] Carnegie Mellon University.

Table II (continued)

Material	Zeta potential in Krebs solution mV	Thrombogenic behavior % occlusion 0%..........100%
Untreated Ethylene-vinyl acetate	—	~15%
Aluminized Ethylene vinyl acetate	—	~20%
Sulfonated Ethylene vinyl acetate	—	~20%
Ethylene vinyl acetate 5% dextran	—	~20%
Untreated Ethylene acrylic acid copolymer	−18.8 to −17.2	~5%
Na$^+$ Ethylene acrylic acid ionomer	−20.7 to −20.1	~5%
Ca^{++} Ethylene acrylic acid ionomer	−14.5 to −16.8	~35%
Mg^{++} Ethylene acrylic acid ionomer	−18.8 to −18.1	~30%
40% DMEA Ethylene acrylic acid	−17.5	~30%
Untreated vinyl acetate Crotonic acid copolymer	—	~10%
Na$^+$ vinyl acetate Crotonic acid ionomer	—	~5%
Texture (paper) Polypropylene	−7.9	~10%
Glow discharged Pyrex glass	−12.2 to −14.8	~3%

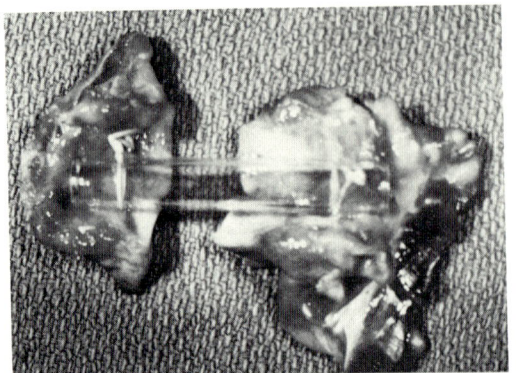

INFLOW P-98-70
    CORNELL AERONAUTICAL LAB. INC.
    PYREX GLASS
    IMPL. 479 DAYS
    SACRIFICED

Fig. 4. The remarkable antithrombogenic behavior of a glow discharge cleaned glass tube implanted in the canine thoracic inferior vena cava for 479 days.

homopolymers as well as their copolymers are negatively charged and inhibit the blood clotting reaction. There are a number of polyelectrolytes which have negative zeta potentials and are antithrombogenic. The most promising is ethylene-acrylic acid copolymer neutralized to the extent of 61% with sodium ions. Another useful related material is the vinyl acetate-crotonic acid copolymer.

Though glass is negatively charged, it is normally thrombogenic. It is very probable that this is due to the heterogeneity of the surface in respect to surface charge. Recently, it was found that glow discharge cleaning of glass [BAIER *et al.*, 1973] makes the material blood-compatible (fig. 4). Untreated glass tubes, implanted in the canine TIVC, are completely occluded in a short time (hours to few days). Similarly, implanted glow discharge cleaned tubes have remained patent for more than 1 year.

IV. Some Recent Techniques for Improving Blood Compatibility

A. Biolized Materials

Biolization of materials of natural or synthetic origin is another approach in the search for blood-compatible materials [IMAI *et al.*, 1971; HOFFMAN *et al.*, 1972]. This process makes use of the fact that, irrespective of its nature, any surface which comes in contact with blood is covered with a protein layer. For these surfaces to obtain a satisfactory degree of antithrombogenicity, it is necessary to cross-link the protein molecules [IMAI et al., 1971]. Biolized natural rubber was prepared by first adding dog albumin, bovine albumin or gelatin solution to prevulcanized natural rubber latex. The modified rubbers were then treated with formaldehyde or gluteraldehyde to effect the cross-linking. In a related procedure, a highly biolized natural rubber was prepared using two biological components, gelatin and heparin. The biolized natural rubbers containing more than 6% protein are superior to silastic, in respect to antithrombogenic properties.

B. Linear or Cross-Linked Homo or Block Polymers

A large number of homo or block elastomers with a desired positive charge concentration as well as of hydrophilic and hydrophobic segments can be obtained by combining any two of the following: (i) polyhydroxy prepolymers reacted with di-isocyanates first and then with dimethyl-amino alcohols; (ii) tetramethyl- or hexamethyl-amino alkanes and α-ω-di-bromo alkanes; and (iii) halo prepolymers formed from di-isocyanates reacted with halo alcohols [RENBAUM et al., 1971]. Heparin can be grafted to these materials by simple impregnation. Only a few of these materials have been prepared and a still lesser number tested for blood compatibility.

C. Surface Pretreatment of Metallic Materials

It is very often necessary to use metallic materials for some types of implants. Surface contamination plays a major role in thrombogenesis of cardiac implants. Electrochemical methods are available to rigorously clean or polish metallic materials. Pulsing and potentiostatic techniques are effective in removing surface contaminants. During this treatment organic films can be oxidized at positive potentials while oxide films can be reduced at cathodic potentials. As smooth surfaces tend to be less thrombogenic than rough surfaces, electropolishing techniques are applicable for practically all metals and alloys. An electropolished aluminum valve, implanted in a calf for 82 days, was found to be strikingly clean (fig. 5) while valves of the same material implanted without pretreatment occluded readily (fig. 6).

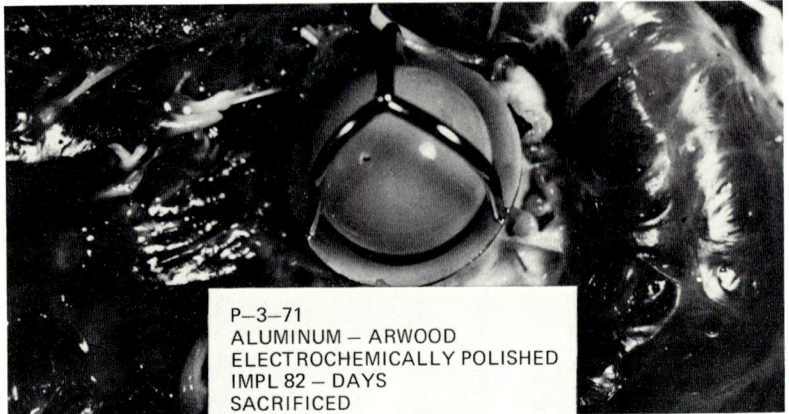

Fig. 5. An electropolished aluminum valve after being implanted in the tricuspid annulus of a calf for 82 days.

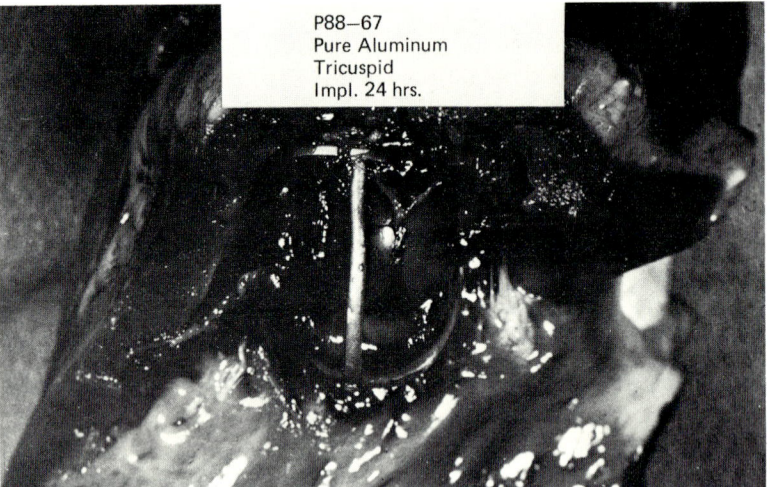

Fig. 6. Occluded aluminum valve implanted 'as received'.

V. Blood-Surface Interactions – Mechanism Studies

A. Général

The criteria for the selection of prosthetic materials for cardiovascular implants and the various techniques of *in vitro* and *in vivo* evaluation of their characteristics have been discussed so far. The utility of an implant is ultimately determined by the long-term patency when exposed to flowing blood stream. The problems of intravascular hemostasis and thrombosis are further complicated by the implants as they themselves can induce or inhibit thrombosis. A study of these blood-surface interactions will lead to a better understanding of the initiation of thrombosis.

Since thrombosis *per se* and the chemical reactions in whole blood are too complex to be useful for basic studies, simpler systems are investigated. The technique of study is dependent on the type of materials used. Thus for insulator materials, i.e., plastics, which are extensively used as implants, the surface interactions can be studied using ellipsometry, electron microscopy and radiotracer techniques. By exposing a large area of the polymers or metals to blood or plasma, the effect on platelets, red cells and coagulability can be investigated. With metallic conductors, many of the electrochemical methods can also be used. Thus the adsorption and electron transfer characteristics of the components of plasma can be investigated.

B. Effect of Metals on the Platelet Release Reaction

Platelets play an important role in hemostasis in the cardiovascular system. When a blood vessel wall is injured, cessation of blood flow from the vascular injury is achieved first by a mass of aggregated platelets. After this, a stable plug is formed by initiation of the blood coagulation cascade leading to the formation of fibrin [HOROWITZ and SPIELVOGEL, 1971].

When prosthetic materials are implanted in the body, they can interfere with the platelet function. A detailed discussion of the platelet release reaction and factors that induce this reaction are outside the scope of this review. When platelets come into contact with a foreign body, usually collagen microfibrils, at the site of an injured blood vessel wall, platelets adhere to the surface. A platelet plug is subsequently formed by aggregation of more platelets at the site of adhesion. Adenosine diphosphate (ADP) is known to induce platelet aggregation [MARCUS and ZUCKER, 1965]. Platelets themselves are the source of the ADP. When platelets aggregate, the intraplatelet constituents are released to the surrounding media. The influence of some metal powders on this platelet release reaction has been studied *in vitro* [ZIVKOVIC *et al.*, 1972]. Washed platelets were treated with some metal powders, and extent of release of adenine nucleotides was studied. The results show that the release reaction is inhibited by Al while noble metals including Pt and Au cause an increased release of adenosine mono-, di- and triphosphates and glucose-

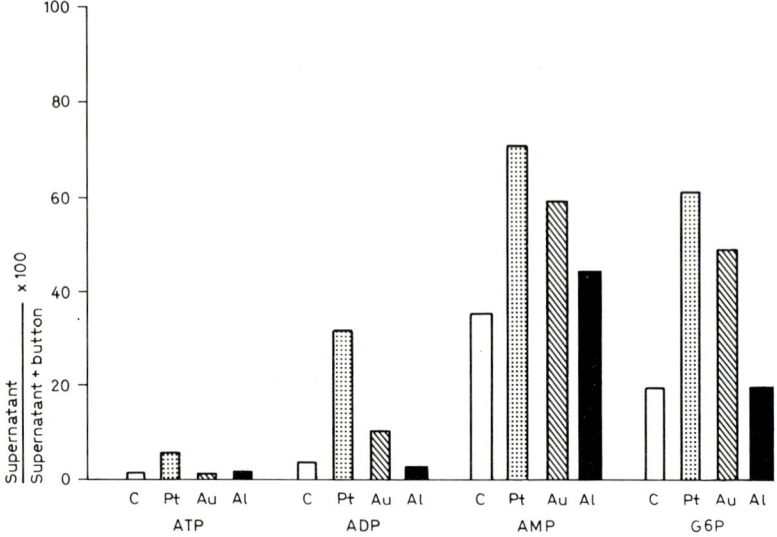

Fig. 7. Release of platelet constituents by exposure of platelets to metal powders. Percentage of adenine nucleotides and glucose-6-phosphate (G6P) in supernatant after centrifugation of platelet suspensions exposed to metal powders.

6-phosphate (fig. 7). The preliminary results with platelet release indicate that metals that are potentially thrombogenic release more nucleotides from platelets. This process is likely to cause platelet aggregation *in vivo* with possible activation of the extrinsic pathway of blood coagulation [MAMMEN, 1971].

C. Electron Microscopic and Optical Studies of
Surfaces Exposed to Biologic Fluids

Electron microscopy has been extensively used to study the nature of deposits on surfaces exposed to biological fluids *in vitro* and *in vivo*. The nature of thrombi deposited on blood vessel walls [WIENER and SPINO, 1962; JORGENSEN, 1965] and on prosthetic materials [STILL *et al.*, 1967; O'NEAL *et al.*, 1964] has been studied using this method. The ultrastructure of thrombi deposited on some prosthetic implants has been investigated using the electron microscope [RANGANATHAN *et al.*, 1970]. The characteristics of thrombi on the various materials studied show significant variations in composition and organization. Scanning electron microscopic techniques are used to study the thrombus formation on

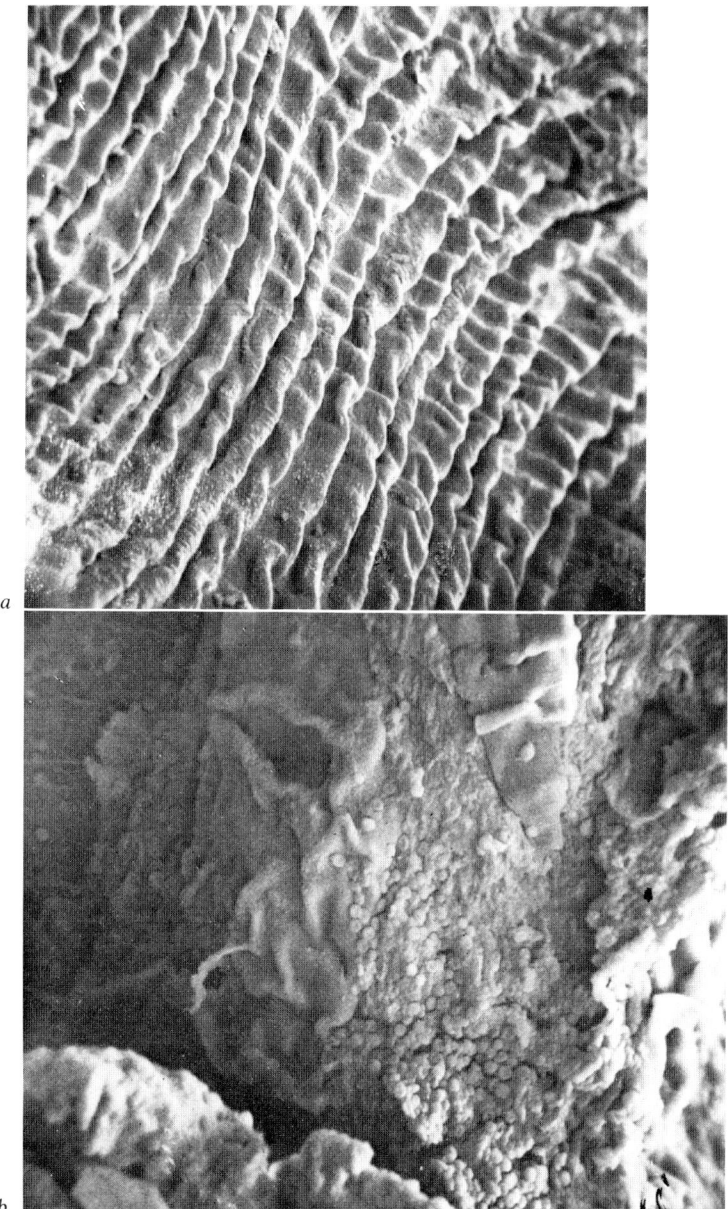

Fig. 8. Scanning electron microscopic photograph of normal endothelium (a) (×840) and thrombus deposited at the junction of the blood vessel wall and prosthetic material (b) (×840).

blood vessels and prosthetic materials. Scanning electron microscope pictures of normal endothelium and a thrombosed blood vessel at the site of prosthetic implant are shown (fig. 8). Scanning electron microscopy and X-ray energy dispersion spectroscopy have been used recently to study the morphology of thrombi deposited on cardiac valves and identify the type of contaminants on prosthetic surfaces [SAWYER *et al.*, 1973, 1974].

Electropolymerization of fibrinogen was investigated using electron microscopy [STONER and WALKER, 1969]. These studies indicate that fibrinogen can be polymerized *in vitro* at anodic potentials, resulting in a product which is morphologically similar to fibrin formed by the action of thrombin on fibrinogen.

Ellipsometry has been used for the study of *in vitro* adsorption of blood proteins on prosthetic materials [VROMAN and ADAMS, 1969a,b; MORRISSEY *et al.*, 1975; RAMASAMY *et al.*, 1973].

D. Electrochemical Reactions of Blood Coagulation Factors

The experiment in which highly thrombogenic copper was made thromboresistant by maintaining the metal at a negative potential shows (section III.A.) that the thrombosis reaction is probably triggered by positive potentials and inhibited by negative potentials. This deduction is further strengthened by the fact that thrombus deposits are observed only on positive electrodes when a current is passed *in vivo* between two metal wire electrodes [SAWYER and SRINIVASAN, 1967]. The weight of thrombus formed is directly proportional to the quantity of charge passed (fig. 9). Differential capacity measurements on gold electrodes immersed in blood show that there is a point of inflection in the capacitance-potential curve at the potential corresponding to the critical potential at which thrombosis is initiated on conducting materials [GILEADI *et al.*, 1969]. All these experiments indicate that the potential at the conducting material-blood interface determines thrombogenesis.

Basically, only two types of reactions can take place at a metal-solution interface, adsorption and charge transfer. Both are potential-dependent phenomena. About nine blood proteins are activated in the blood coagulation cascade, resulting in the conversion of prothrombin to thrombin which polymerizes fibrinogen to fibrin. Individual purified blood proteins suspended in saline (0.154 M NaCl) were used for the investigations. Cyclic voltametry and steady state potentiostatic studies have been carried out with prothrombin [DUIC *et al.*, 1973], fibrinogen [RAMASAMY *et al.*, 1973], factors V and VIII [Ramasamy *et al.*, 1974a,b].

The proteins investigated so far show distinct cyclic voltametric peaks

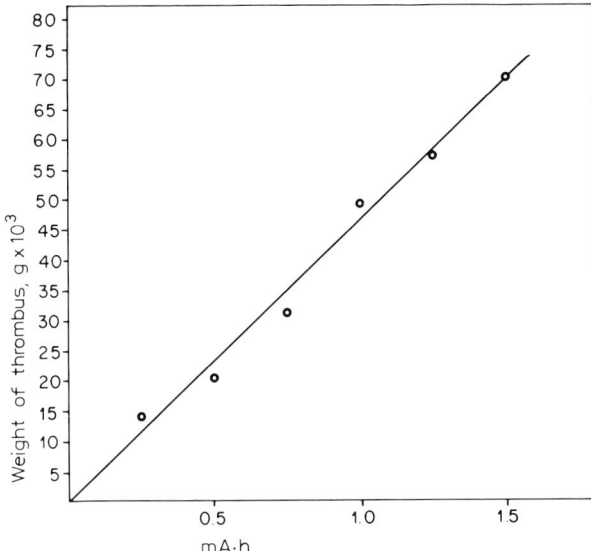

Fig. 9. Relation between quantity of electricity passed and the weight of thrombus deposited.

at cathodic potentials [more negative than 600 mV/SCE (standard calomel electrode)]. The reactions characterized by these peaks appear to be significantly influenced by adsorption. This is revealed by application of some of the diagnostic criteria for the influence of adsorption and charge transfer reactions to the cathodic peaks of cyclic voltamograms [WOPSCHALL and SHAIN, 1967; NICHOLSON and SHAIN, 1964].

Constant potential studies with fibrinogen indicate that the protein possibly takes part in two charge transfer reactions, one at cathodic potentials and the other over a wide range of anodic potentials [RAMASAMY *et al.*, 1973; STONER and WALKER, 1969]. The electrode product of anodic reaction has been electron-microscopically examined. This was shown to be similar to the fibrin formed by thrombin addition to fibrinogen. Recently, fibrinogen solutions subjected to cathodic and anodic potentiostatic treatments were studied for biologic activity. An extension of coagulation time (thrombin time) was observed for the protein maintained at cathodic potentials while a shortening of coagulation time results when the protein is maintained at anodic potentials [RAMASAMY *et al.*, 1973]. There is no significant decrease in fibrinogen level as determined spectrophotometrically in both reactions. Recent studies, however, indicate that these changes in clotting times could be due to changes in pH near the metal-protein solution interface [RAMASAMY *et al.*, 1975]. Potentiostatic treatment of prothrombin at anodic potentials (> +1,000 mV/SCE) results in generation of a product that has thrombin-like activity [DUIC *et al.*, 1973]. The results are preliminary. Further studies are in progress to quantify the findings.

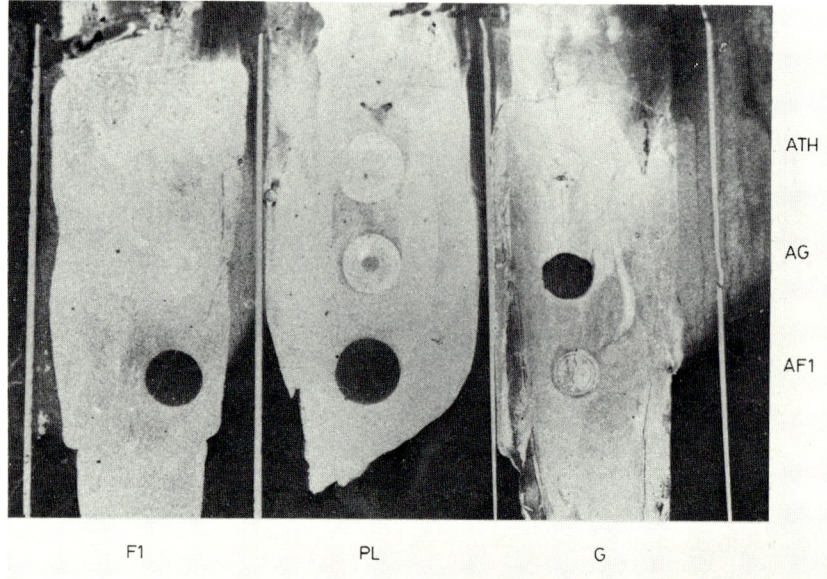

Fig. 10. Condensed water vapour patterns on glass that had adsorbed films by exposure to fibrinogen (Fl) for 1 min, normal intact citrated plasma (PL) for 5 sec and 7S γ-globulin (G) for 1 min. On each of these, the following antisera were added: antithyroglobulin (ATH), anti-immunoglobulin (AG) and antifibrinogen (AFl). They were rinsed after 1 min, then dried and exposed to water vapor. Water-wettable areas were left by AG on globulin film and by AFl on both fibrinogen and plasma film.

Fig. 11. a Effect of intact (int.) and activated (act.) plasma and albumin. *b* 7S γ-globulins and albumin upon preabsorbed fibrinogen films. Ellipsometer recordings. An oxidized Si slide was used as the reflecting solid. One recorder unit is equivalent to about 1 A. Human fibrinogen was adsorbed out of 15 ml of veronal-buffered saline to which 1.2 mg of plasminogen-free fibrinogen was added at time 0. 60 min later, solutions were replaced by 15 ml of fresh buffer, recordings were taken and 0.1 ml of plasma or 0.2 ml (0.4 mg) of crystalline human albumin or 7S γ-globulins was added. 45 min later the contents of the cuvette were replaced again by buffer, readings were taken and 0.1 ml of rabbit anti-human fibrinogen was added.

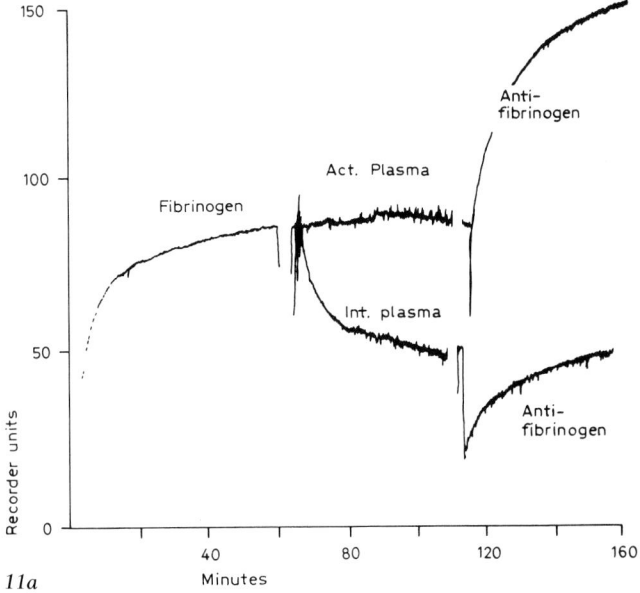

11a

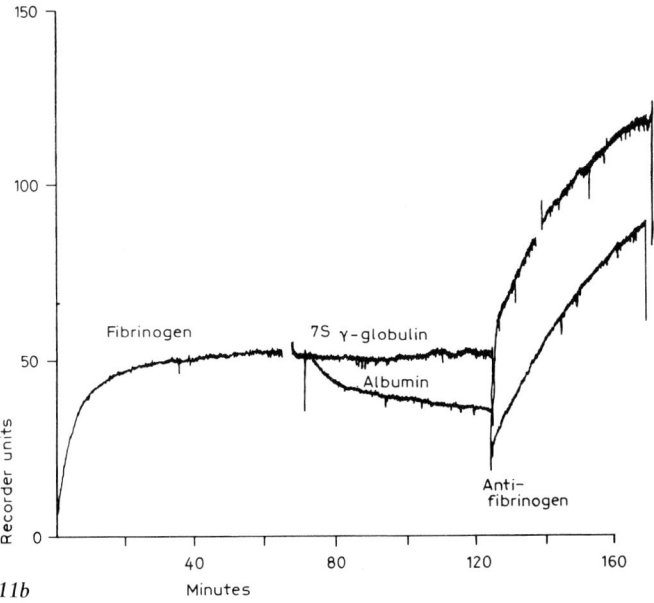

11b

E. Adsorption and Adsorption Inhibition

The adsorption of blood proteins on metallic materials and its dependence on potential can be investigated by a number of methods. Potential-dependent adsorption of thrombin and fibrinogen on platinum has been studied using ellipsometry [RANGANATHAN et al., 1970]. Differential capacity and electrocapillary methods were also employed to study the adsorption of fibrinogen, thrombin and Hageman factor [STONER, 1969; STONER and SRINIVASAN, 1970]. Irrespective of the technique employed, fibrinogen always shows strong adsorption at metal-solution interfaces over a wide potential range. It is adsorbed even at very low concentrations. This observation agrees well with similar behavior observed on insulating materials [VROMAN and ADAMS, 1969a,b]. At highly cathodic potentials, desorption takes place while the coverage increases with increase of anodic potentials. The reason for the strong adsorption of fibrinogen is probably the strong π bond interactions of the tryptophan residue of the protein with the metal.

Hageman factor is believed to trigger the intrinsic blood coagulation scheme. It is interesting to note that Hageman factor is not adsorbed at the mercury-solution interface, as revealed by electrocapillary measurements [STONER and SRINIVASAN, 1970].

A rapid method for the study of adsorption of proteins on insulator materials was developed [VROMAN and ADAMS, 1971]. It is based on the hydrophobicity of protein layers and interaction of antisera with their matching proteins. Using this method, it was shown that fibrinogen deposits within 5 sec onto a glass plate immersed in plasma. The reactions appear to be far more complex with preadsorbed protein films, which are treated with human plasma followed by specific antisera (fig. 10). The technique of recording ellipsometry was also used for these studies [VROMAN, 1971]. Fibrinogen films on oxidized Si crystal slices lose thickness on exposure to all proteins, antisera and plasma (fig. 11).

This case of adsorption of fibrinogen may have serious implications *in vivo*. Clotting may not be initiated by the adsorption, though the surface can act as a site for platelets to adhere. It is well known that for platelet adhesion to glass, fibrinogen is necessary [GUGLER and LUSCHER, 1965; HATHAWAY et al., 1965; MUSTARD et al., 1967]. Thus the first step preceding platelet aggregation (i.e., adhesion) may be triggered by the adsorbed fibrinogen on surfaces.

The adsorption of fibrinogen is influenced by heparin. Heparin is a naturally occurring anticoagulant in mammals. It is interesting to note that prior adsorption of heparin on mica surfaces inhibits adsorption of fibrinogen. The highly negatively charged heparin units, adsorbed on the surface, prevent the subsequent adsorption of fibrinogen [Stoner and Srinivasan, 1970]. Heparinized polymers do not show similar behavior towards adsorption of fibrinogen. Heparinization of one type of aminated polymer prevents the adsorption of proteins, while it has no effect on another aminated polymer [Vroman and Adams, 1969a].

VI. Conclusions

Thrombosis is an interfacial reaction occurring at a solid-electrolyte interface. It has been suggested that the surface charge of a material is an important parameter in determining thrombogenesis. Another important property which has been suggested is critical surface tension. Conducting materials registering potentials below +100 mV in blood versus the potential of the NHE tend to be nonthrombogenic. The corrodible metals are in this category. There are some passive alloys which have good blood-compatible characteristics. Surface pretreatment to remove oxide and organic films is very often essential prior to biological testing. Insulator materials with a uniform negative charge tend to be nonthrombogenic. Chemical treatment of insulator materials to introduce negatively charged groups such as sulfonate, carboxylate, etc. enhances their blood compatibility. Electrets with a negative surface charge are more resistant to thrombus deposition than the corresponding untreated materials. Heparinization of polymers (silicone rubber, epoxy) has been attempted but in most cases the heparin is bonded ionically and is not stable in physiological solutions. Cross-linking the heparin on the surfaces with gluteraldehyde prevents it from elution. There are some polyelectrolyte materials (ethylene-acrylic acid and vinyl acetate-crotonic acid copolymers and their partially neutralized salts) which are nonthrombogenic. Glow discharged glass tubes implanted in the TIVC have remained patent for more than 1 year.

Recently, attempts have been made to bond proteins to surfaces. Enzyme-coated materials may prove to be quite useful. At the present time, the most promising blood-compatible materials are aluminum, titanium, Stellite 21, a proprietary cobalt-chromium alloy, ethylene-acrylic acid copolymer 61% neutralized with sodium ions, sulfonated Teflon and some hydrogels.

The morphology of the initial layers of thrombus deposits depends on the type of prosthetic material to which blood is exposed. Adsorption and competitive adsorption reactions at the prosthetic material-blood/plasma interface seem to be very complex. Adsorption of individual blood proteins on metallic prostheses depends on the potential. Fibrinogen is almost invariably the first protein to be adsorbed on any surface from plasma. It may be 'converted' on standing and may be influenced by the presence of factor XII in the medium.

The presence of anticoagulants like heparin tends to inhibit the adsorption of fibrinogen. A few of the blood coagulation proteins appear to undergo charge transfer reactions at the metal-solution interface. No quantitative information is available yet regarding the nature of these reactions. It is possible that the reactions interfere with the biological activity of blood coagulation proteins. In all probability, similar reactions take place at the implanted material-blood interface *in vivo*.

Acknowledgements

Most of the work reported in this article was supported by grants (HE 07371, HE 07821 and HE 10795) and a contract (PH 43-68-75) from the National Heart and Lung Institute, National Institutes of Health, Bethesda, Md. Small grants for parts of the experimental work were also provided by American Cyanamid Company and Howmedica, Inc. The authors also wish to express their deep appreciation to their former or present associates Drs. J. G. MARTIN, P. S. CHOPRA, M. E. LEE, T. MURAKAMI, A. PARMEGGIANI, R. RUBIN, R. V. ZIVKOVIC, L. DUIC, E. GILEADI, G. E. STONER, and Messrs. R. MORRISON, T. R. LUCAS, and C. B. BURROWES for their contributions to the work. One of us (Dr. S. SRINIVASAN) was on a Career Scientist Award from the Health Research Council, City of New York, during the period 1967–1972 (contract I 542).

References

BAIER, R. E.; DEPALMA, V. A.; FURUSE, A.; GOTT, V. L.; LUCAS, T. R.; SAWYER, P. N.; SRINIVASAN, S., and STANCZEWSKI, B.: Thromboresistance of glass cleaned by glow discharge treatment in argon. Am. Soc. artif. internal Organs Abstr. *2:* 3 (1973).
BAIER, R. E. and DUTTON, R. C.: Initial events in interactions of blood with a foreign surface. J. Biomed. Mater. Res. *3:* 181 (1969).
CHOPRA, P. S.; SRINIVASAN, S.; LUCAS, T. R., and SAWYER, P. N.: Relation between thrombosis on metal electrodes and the position of metal in the electromotive series. Nature, Lond. *215:* 1494 (1967).
DUIC, L.; SRINIVASAN, S., and SAWYER, P. N.: Electrochemical behavior of blood coagulation factors: prothrombin and thrombin. J. electrochem. Soc. *120:* 348 (1973).
EDMARK, K. W.; DAVIS, J., and MILLIGAN, H. L.: Streaming potential – an analysis of the test technique and its application to the study of thrombosis on implants. Thromb. Diath. haemorrh. *24:* 286 (1970).
GILEADI, E.; SRINIVASAN, S., and SAWYER, P. N.: Electrochemical methods for the determination of thrombus formation on conducting surfaces, J. electroanal. Chem. Interf. Electrochem. *21:* 6 (1969).
GILEADI, E.; STANCZEWSKI, B.; PARMEGGIANI, A.; LUCAS, T. R.; RANGANATHAN, M.; SRINIVASAN, S., and SAWYER, P. N.: Antithrombogenic characteristics of cathodically polarized copper prostheses. J. Biomed. Mater. Res. *6:* 489 (1972).

GORTNER, R. A. and BRIGGS, D. R.: Glass surfaces versus paraffin surfaces in blood clotting phenomena – a hypothesis. Proc. Soc. exp. Biol. Med. 25: 820 (1928).

GRODE, G. A.; FALB, R. D., and ANDERSON, S. J.: Development of materials for use in circulatory assist devices; in HEGYELI (ed.) Artificial Heart Program Conf. Proc., Washington 1969 (US Government Printing Office, Washington 1969).

GUGLER, E. and LUSCHER, E. J.: Platelet function in congenital afibrogenemia. Thromb. Diath. haemorrh. 14: 361 (1965).

HATHAWAY, W. E.; BELHASEN, L. P., and HATHAWAY, H. S.: Evidence for a new plasma thromboplastin factor. I. Case report, coagulation studies and physicochemical properties. Blood 26: 521 (1965).

HOFFMAN, A. S.; SCHMER, A.; HARRIS, C., and KRAFT, W. A.: Covalent binding of biomolecules to radiation grafted hydrogels on inert polymer surfaces. Trans. Am. Soc. artif. internal Organs 18: 10 (1972).

HOROWITZ, H. I. and SPIELVOGEL, A. R.: Haemostasis; in BANG, BELLER, DEUTSCH and MAMMEN Thrombosis and bleeding disorders, chapt. 9 (Academic Press, New York 1971).

IMAI, Y.; TAKIMA, K., and NOSE, Y.: Biolized materials for cardiovascular prostheses. Trans. Am. Soc. artif. internal Organs 17: 6 (1971).

JORGENSEN, L.: Fate of platelet mural thrombi in swine arteries. Circulation 32: suppl. 11, p. 19 (1965).

LAGERGREN, H. R. and ERICKSSON, J. C.: Plastics with a stable surface monolayer of cross-linked heparin: preparation and evaluation. Trans. Am. Soc. artif. internal Organs 17: 10 (1971).

LEE, M. E.; MURAKAMI, T.; PARMEGGIANI, A., and SRINIVASAN, S.: Etiology of thrombus formation on prosthetic metal heart valves. J. thorac. cardiovasc. Surg. 63: 809 (1972).

LEININGER, R. I.; EPSTEIN, M. M.; FALB, R. D., and GRODE, G. A.: Preparation of nonthrombogenic plastic surfaces. Trans. Am. Soc. artif. internal Organs 12: 151 (1966).

LEININGER, R. I.; MIRKOVITCH, V.; BECK, R. E.; ARDUS, P. G., and KOLFF, W. J.: The zeta potentials of some selected solids in respect to plasma and plasma fractions. Trans. Am. Soc. artif. internal Organs 10: 237 (1964).

LYMAN, D. J.; BRASH, J. L., and KLEIN, K. G.: The effect of chemical structure and surface properties of synthetic polymers on the coagulation of blood; in HEGYELI (ed.) Artificial Heart Program Conf. Proc., Washington 1969, (US Government Printing Office, Washington 1969).

MAMMEN, E. F.: Physiology and biochemistry of blood coagulation; in BANG, BELLER, DEUTSCH and MAMMEN. Thrombosis and bleeding disorders, chapt. 1 (Academic Press, New York 1971).

MARCUS, A. J. and ZUCKER, M. B.: The physiology of blood platelets (Grune & Stratton, New York 1965).

MILLIGAN, H. L.; DAVIS, J., and EDMARK, K. W.: The search for correlations between electrokinetic phenomena and blood thrombus formation on materials. J. Biomed. Mater. Res. 2: 51 (1966).

MORRISSEY, B. W.; SMITH, L. E.; FENSTERMAKER, C. A., and STROMBERG, R. R.: Dependence of adsorbed blood protein conformation on surface potential. 147th Meet. Electrochem. Soc., Toronto. Extended Abstr. No. 383 (1975).

MURPHY, P.; CACROIX, A., and MERCHANT, S.: Studies relative to materials suitable for use in artificial hearts; in HEGYELI (ed.) Artificial Heart Program Conf. Proc., Washington 1969, chapt. 10 (US Government Printing Office, Washington (1969).

MUSTARD, J. F.; GLYNN, M. F.; NISHZAWA, E. E., and PACKHAN, M. A.: Platelet surface interactions: relationship to thrombosis and hemostasis. Fed. Proc. 26: 106 (1967).

NICHOLSON, R. S. and SHAIN, I.: Theory of stationary electrode polarography – single scan and cyclic methods applied to reversible irreversible and kinetic systems. Analyt. Chem. 36: 706 (1964).

O'NEAL, R. M.; JORDAN, G. L., jr.; DE BAKEY, M. E., and HALPERT, B.: Cells grown on isolated intravascular dacron hubs; an electron microscopic study. Expl. molec. Path. 3: 403 (1964).

PETSCHEK, H. E. and MADRAS, P. N.: Thrombus formation on artificial surfaces; in HEGYELI (ed.) Artificial Heart Program Cong. Proc., Washington 1969, (US Government Printing Office, Washington 1969).

POORE, G. V.: A text book of electricity in medicine and surgery (Appleton, New York 1876).

RAMASAMY, N.; KEATES, J. S.; SRINIVASAN, S., and SAWYER, P. N.: Electrochemical and enzymatic behavior of fibrinogen. Bioelectrochem. Bioenerg. 1: 244 (1974a).

RAMASAMY, N.; LUCAS, L.; KEATES, J. S.; SRINIVASAN, S., and SAWYER, P. N.: Fibrinogen coagulation kinetics – effects of metal powders and pH. Unpublished results (1975).

RAMASAMY, N.; RANGANATHAN, M.; DUIC, L.; SRINIVASAN, S., and SAWYER, P. N.: Electrochemical behavior of blood coagulation factors: fibrinogen. J. Electrochem. Soc. 120: 354 (1973).

RAMASAMY, N.; SRINIVASAN, S., and SAWYER, P. N.: Electrochemical behavior of blood coagulation proteins – factors V and VIII. Electrochim. Acta 19: 137 (1974b).

RANGANATHAN, M.; STEMPAK, J. G.; SRINIVASAN, S., and SAWYER, P. N.: Junctional thrombi deposited on conducting materials at different potentials – an electron microscopic study. Thromb. Diath. haemorrh. 24: 273 (1970).

RENBAUM, A.; YEN, S. P. S.; LANDEL, R. F., and SHEN, M.: Synthesis and properties of a new class of potential biomedical polymers; in RENBAUM and SHEN Biomedical polymers, p. 221 (Marcel Dekker, New York 1971).

SAWYER, P. N. and SRINIVASAN, S.: Studies on biophysics of intravascular thrombosis. Am. J. Surg. 113: 42 (1967).

SAWYER, P. N.; STANCZEWSKI, B.; RAMASAMY, N.; KAMMLOTT, G. W.; STEMPAK, J. G., and SRINIVASAN, S.: Electrochemical and chemical methods for production of nonthrombogenic metal heart valves: combined biophysical, electron microscopic and scanning electron microscopic studies. Trans. Am. Soc. artif. internal Organs 19: 195 (1973).

SAWYER, P. N.; STANCZEWSKI, B.; SRINIVASAN, S.; STEMPAK, J. G., and KAMMLOTT, G. W.: Electron microscopy and physical chemistry of healing in prosthetic heart valves, skirts and struts. J. thorac. cardiovasc. Surg. 67: 24 (1974).

SCUDAMORE, C.: Essay on blood (Longham, Hurst, Rees, Orme, Brown & Green, London 1824).

SKENE, A.: Electrical hemostasis in operative surgery (Appleton, New York 1899).

SRINIVASAN, S.; COHEN, J. L., and SAWYER, P. N.: Preliminary studies on the electrode kinetic mechanism of thrombosis. Proc. Int. Union Physiol. Sciences, Abstr. VII, p. 414 (1968).

SRINIVASAN, S.; DUIC, L.; RAMASAMY, N.; SAWYER, P. N., and STONER, G. E.: Electrochemical reactions of blood coagulation factors – their role in thrombosis. Beri. Bunsen-Ges. phys. Chem. 77: 798 (1973).

SRINIVASAN, S. and SAWYER, P. N.: Electrochemical techniques in studies on intravascular thrombosis. J. Ass. Advancement med. Instrumentation 3: 116 (1969).

SRINIVASAN, S. and SAWYER, P. N.: Role of surface charge of the blood vessel wall, blood cells and prosthetic materials in intravascular thrombosis. J. Coll. Interf. Sci. *32:* 456 (1970).

STILL, W. J. S.; GHANI, A. R., and DENNISAN, S. M.: The organization of isolated mural thrombi in aortic grafts – an electron microscopic study. Am. J. Path. *51:* 1013 (1967).

STONER, G. E.: Electrosorption of amino acids, peptides and proteins in relation to the compatibility of materials and the human body. J. Biomed. Mater. Res. *3:* 655 (1969).

STONER, G. E. and SRINIVASAN, S.: Adsorption of blood proteins on metals using capacitance techniques. J. phys. Chem. *74:* 1088 (1970).

STONER, G. E.; SRINIVASAN, S., and GILEADI, E.: Adsorption inhibition as a mechanism for the antithrombogenic activity of some drugs. I. Competitive adsorption of fibrinogen and heparin on mica. J. phys. Chem. *75:* 2107 (1971).

STONER, G. E. and WALKER, L.: The enzymatic and electropolymerization of fibrinogen. J. Biomed. Mater. Res. *3:* 645 (1969).

VROMAN, L.: Interactions among human blood proteins at interfaces. Fed. Proc. *30:* 5 (1971).

VROMAN, L. and ADAMS, A. L.: Effect of heparin on reactions at aminated polymer-blood interfaces. J. Coll. Interf. Sci. *31:* 188 (1969a).

VROMAN, L. and ADAMS, A. L.: Findings with the recording ellipsometer suggesting rapid exchange of specific plasma proteins at liquid solid interfaces. Surface Sci. *16:* 438 (1969b).

VROMAN, L. and ADAMS, A. L.: Identification of adsorbed protein films by exposure to antisera and water vapors. J. Biomed. Mater. Res. *3:* 669 (1971).

WIENER, J. and SPINO, D.: Electron microscopic studies in experimental thrombosis. Expl. molec. Path. *1:* 558 (1962).

WOOD, L. A.; HORAN, F. E.; SHEPPARD, E., and WRIGHT, I. S.: Zeta potential measurements as a tool for studying certain aspects of blood coagulation; in FLYNN Blood clotting and allied problems (Josiah Macy Foundation, New York 1950).

WOPSCHALL, R. H. and SHAIN, I.: Effect of adsorption of electroactive species in stationary electrode polarography. Analyt. Chem. *39:* 1514 (1967).

ZIVKOVIC, R. V.; SRINIVASAN, S., and SAWYER, P. N.: Effect of some metals on the platelet release reaction. Abstr. Int. Soc. Thrombosis Haemostasis, IIIrd Congr., Washington 1972, p. 266.

Dr. S. SRINIVASAN, Department of Energy and Environment, Brookhaven National Laboratory, *Upton,* NY 11973 (USA).

Subject Index

Activation factor 114, 116–118, 124, 125
Adenosine triphosphate 151
Adsorption 158
albumin 121, 149
Alveolar-capillary membrane 49
Alveoli 49
Anticoagulant 134
Antithrombogenic materials 137, 140, 144–147
Antithrombotic 134, 136
Artificial heart-lung machine 72

Biocompatibility 107, 108
Biolization 148
Biolized materials 148
Block elastomer 149
Blood coagulation 115–117, 123, 154
– – factors 118, 154
– compatibility of materials, criteria 135
– trauma 73
– –, denaturation 74
– –, hemolysis 73, 75
Blood-surface interactions 150
Bubble dynamics 79
– –, formation or cavitation 80
– –, growth or shrinkage 82
– –, multi-gas diffusion 96
– –, single-gas diffusion 95
– lifetime 65

Caisson equation 12
Capacitance-potential curve 154
Carboxylate groups 143
Coagulant 134

Coagulum 134
Collagen microfibrils 151
Concentration vs. time for dissolved gas in blood 63
Conducting materials 137, 140
Contact angle equation 103
Copolymers 149
Corrodible metals 140
Couette viscometer 8, 9, 15
Coulter counter 29
Covalent bonding 143
Cross-linking of heparin 143

Decompression sickness 82
– –, prevention 83
– –, symptoms 82
Degassing process apparatus 68
Denaturation 74
Diffusing capacity 50
– –, lung 53
Diffusion, carbondioxide 50
–, oxygen 50
Dissolution, carbondioxide in blood 68
–, gas bubble in blood 67
–, gas eboli in tissues 86
– – – –, tests 87
– – – –, theory 91
–, nitrogen in blood 68
–, oxygen in blood 68
Dynamics, gas cavity in tissue-capillary 91

Electrochemical reactions 154
Electroconduction 125
Electrokinetic 143

Subject Index

Electron microscopy 151
Electroosmosis 143
Electropolishing 149
Electropolymerization 154
Ellipsometry 110, 151, 154
Epoxy 143
Erythrocyte flexibility 2

Fahraeus-Lindquist effect 2, 21
Fibrinogen 121, 154, 155, 158
Fibrinogen-platelet interaction 122
Fick's law of diffusion 50
Foreign materials 119
Formaldehyde 149
Free surface 101
Freundlich equation 109

Gas embolism 57
– –, behavior controlled by liquid inertia 65
– –, dissolution in tissues 86
– –, due to decompression 57, 80
– –, due to extracorporeal blood oxygenation 58, 72
– –, effect of foreign agents (plasma substitutes and anesthetics) 70, 71
– –, following hypothermia 58
– –, heart-lung machine 73
– –, introduction and prevention during open-heart surgery 74–77
– –, sources 57
– –, theory of expansion and dissolution in blood 59
– exchange 49
– – between alveolar air and capillary blood 52
Gibbs free energy 120
Globulin 121
Gluteraldehyde 143, 149

Haematocrit 29
–, centrifuge 29
Hageman factor 158
Hagen-Poiseuille law 19
Helmholtz free energy 102
Hemolysis 73
–, basis 75
Hemostasis 116, 134
Heparinized polymers 159
Homo elastomer 149
– polymers 149

Hydrophobic surfaces 112, 113
Hydrophobicity 158
Hydroxyl groups 143

Implantable materials 137, 140, 144, 147
Insulator materials 143
Interface, between foreign materials and blood 101, 103
–, solid-liquid 105
Interfacial electrochemical property 136
– parameters of materials 107
– reaction 134
Intravascular hemostasis 151
Isoelectric point–113

Langmuir equation 109

Maxwell model 22
Metallic materials 149
Minimal interfacial free energy 108

Non-Newtonian fluids 2

Oxygen uptake rate in lung capillaries 55
Oxygen-hemoglobin dissociation curve 55

Partial pressures, respiratory gases 49, 51
Platelet adhesion 136
– – and aggregation 115
– – on foreign materials 119
– aggregation 134
– release reaction 151
Polyelectrolyte materials 159
Polymer surfaces 107
Polyurethane value 142
Porous materials 143
Potentials, anodic and cathodic 158
– of implanted valves 140
Potentiostatic 156
Prepolymer 149
Protein, absorption 109
Protein adsorption 108
– –, foreign surfaces 109
– –, measuring techniques 109
–, denaturation 113
–, interface interactions 110
– – –, apolar surfaces 111
– – –, polar surfaces 111
Prothrombin 156
Pulmonary or venous embolism 78

Subject Index

Pyrolytic carbon 143

Radius versus time for dissolving bubbles in blood 63, 64
– – – of CO_2 in degassed plasma 69
– – – of O_2 bubbles in degassed plasma 69
Relaxation time 2
Rheology, blood following heart infarction 31–34
– – plasma 7–11
–, normal blood 11–23, 29, 30
–, viscoelastic fluids 3
Rheometer 28

Saccharide chains 122
Semiconduction 125
Shear rate 2, 5
– stress 2, 4
Silicon rubber 124, 143
Streaming potential 143
Surface energy, 101, 105, 136
– parameters of materials 107
– tension 102, 105
– –, critical 104
– –, solid-liquid system 104–106
Stellite valve 142

Teflon 143
Thrombogenesis 159
Thrombogenic behavior, conducting materials 137–140
– –, insulating materials 144, 145
Thrombosis 135
Tissue creep 96
Tissue-capillary gas exchange 84
– – –, carbon dioxide 85
– – –, oxygen 85
Titanium 141

Viscoelastic fluids 2
Viscometer, capillary 21–27
–, cone-plate 12, 18, 22
–, Couette system (with a narrow gap) 8, 9, 22
– – – (with a wide gap) 14, 15, 22
–, sphere-sphere system 8–11
Viscosity 4, 5
–, apparent 23
–, blood, affected by smoking 30, 36, 39
– –, anaemia 38
– –, diabetes 35
– –, disease states 35–38
– –, following heart infarction 31–33
– –, peripheral vascular disease 39
– –, plasmocytoma 37
– –, rheumatic disease 37
– –, under medications to reduce it 34
–, clinical pathology and medicine 24
–, normal human blood 28
–, plasma, effect on circulation 34
– – (normal and following infarction) 31
–, shear 15
–, variable 6, 14, 16
–, zero 5

Water wettability 107
Work of adhesion, 103, 107

Yield shear stress 2, 8, 12
Young-Dupree equation 103

Zeta potential 134, 135

RAYMOND H. FOGLER LIBRARY

DATE DUE